Integrative Neurology:

Hope and Help for People Suffering From Neuropathy

By

George W. Kukurin DC DACAN
Chiropractic Neurologist

*There is nothing covered, that shall not be revealed;
and hid, that shall not be known. -- Bible, 'Matthew' 10:26.*

Hope and Help for People Suffering From Peripheral Neuropathy

Dr. George W Kukurin

www.kcpv.info

KCN Press

Pittsburgh, Pennsylvania ● Phoenix, Arizona

IBSN: 978-0-6152-2065-9

TABLE OF CONTENTS

About The Author

Dr. Kukurin has Post-Graduate education from some of the world's finest institutions, including Johns Hopkins, Harvard Medical School, and the Mayo Clinic. The knowledge he acquired from these world-class medical centers has helped him to develop techniques for his patients that produce outstanding results. The results his patients have obtained are so remarkable that they have been presented to other doctors at Johns Hopkins, appear in Medical Journals indexed with the National Library of Medicine, and have been featured on local NBC, ABC & Fox News affiliates.

Dr. Kukurin has had a lifelong interest in neurology and is a member of the charter class of Chiropractic Neurologists. He scored a perfect 100% on the clinical portion of the Chiropractic Neurology Board Examinations, leading to Diplomat Status with the American Chiropractic Academy of Neurology. He has taught neurology on a post graduate level for the Parker College in Dallas, Texas, making him ideally suited to manage difficult neurological conditions like peripheral neuropathy, spinal stenosis, neuralgias, migraines, and other complex conditions of the nervous system. He recently has authored several books on his methods for helping patients with various difficult conditions of the nervous system.

Dr Kukurin has more than 20 years experience combining acupuncture, chiropractic, herbal, and other integrative medicine techniques into a safe, effective system proven to produce dramatic results for his patients.

Various **Who's Who Organizations** and the **Consumers Research Council of America** have recognized Dr. Kukurin, the remarkable techniques he has developed for his patients, and the results his patients obtain by listing Dr. Kukurin among the *Top Doctors in America.*

Section I: The physiopathology and treatment of neuropathy using integrative neurology methods.

What is neuropathy?

The literal translation of neuropathy means: neuro = pertaining to the nerves, and pathy = sickness. Neuropathy then means "sick nerves". When nerves get sick, people have big problems. The nervous system controls and coordinates all functions of the body. It is the communication system that carries information from one part of the body to another and allows organs to communicate and react to changes both inside the body and in the outside environment in which the body resides.

When nerves are damaged, they either send signals when they are not supposed to or don't send signals when they should. This plays havoc with the complex command and control systems of the body by knocking out proper communications between various organs and tissues.

Anything that interferes with the normal communications within the body has far-reaching negative affects on a person. To put it simply, sick nerves mean a sick person.

Neuropathy can occur anywhere in the body, but is common in the lower extremities, in particular, the feet. Often the feet are affected first and the symptoms progress up the body. The hands are also commonly affected. Clinically, a classic sign of neuropathy is pain and/or numbness in a "glove and stocking" pattern, meaning the patient has troubles in their hands and feet. In severe cases of neuropathy, the nerves that control the stomach, heart, or other organs can become damaged. This causes all sorts of secondary signs and symptoms, depending on which nerves are damaged and the organs that are controlled by those nerves.

Even when the nerves are starting to regenerate, troublesome symptoms may appear. As nerves are damaged, they repair by sprouting (see Figure 1). These immature nerve sprouts, called rootlets, are very sensitive. They send signals spontaneously and are very sensitive to stimulation. This hypersensitivity is termed allodynia. Even a patient's clothes touching the skin in the area of these immature rootlets can cause pain. For a patient to obtain good long lasting relief, it is essential to help these immature rootlets develop and reduce their mechanical sensitivity. We incorporate ultrasound and laser treatments which have been shown to help in this regard. The utility of these and other therapies are discussed in detail later in this paper.

Neuropathy is one of the most common afflictions of the nervous system. Despite the fact that the condition affects so many people, the typical approach to treatment is usually inadequate. At the time I wrote this book, there were only two drugs the FDA approved for the treatment of neuropathy. The lack of effective treatment for this painful condition is one of the reasons why I wrote this book.

Diagnosing Neuropathy

Neuropathy can usually be diagnosed from a good history and physical examination. To confirm the diagnosis, nerve tests known as nerve conduction velocity studies (NCVs) and electromyography (EMGs) are often performed on patients. The diagnosis of neuropathy probably does not require these advanced diagnostic tests, but they do provide information about the severity of neuropathy. These tests are very sensitive and can demonstrate changes in the nervous system long before a patient realizes they have neuropathy.

Figure 1, an electron micrograph showing immature nerve rootlets sprouting from a damaged neuron (nerve cell). These rootlets are extremely sensitive, even sending pain signals spontaneously. This means that they may send pain signals to the brain even when no tissue destruction is occurring. They may also send pain signals to the brain in the presence of a non-painful stimulus, like vibration, light touch, or even clothes touching the skin. If these pain-sensitive rootlets mature and reconnect to their original target tissues, they lose much of their mechanical sensitivity. This reduces pain for the neuropathy patient. Helping these rootlets to mature should be a goal of any neuropathy treatment program.

Likewise, they can also document positive changes in a patient before he or she can "feel" a difference in their condition. So they can be used to monitor response to treatment. They can be useful in providing objective evidence that a treatment is succeeding, or they may indicate that another approach to treatment is warranted.

However, as an alternative to these expensive tests, a skilled doctor using a tuning fork, straight pin, Q-Tip, and reflex hammer can usually diagnose a patient suffering from neuropathy. It is important, in fact mandatory, to have blood work done to try to determine the cause of the neuropathy. It makes no sense trying to heal damaged nerves until the reason for the nerve damage is first eliminated.

A very early sign of neuropathy is loss of the ability to feel the sensation of vibration in the toes. This can be tested quite easily using a simple tuning fork, figure2.

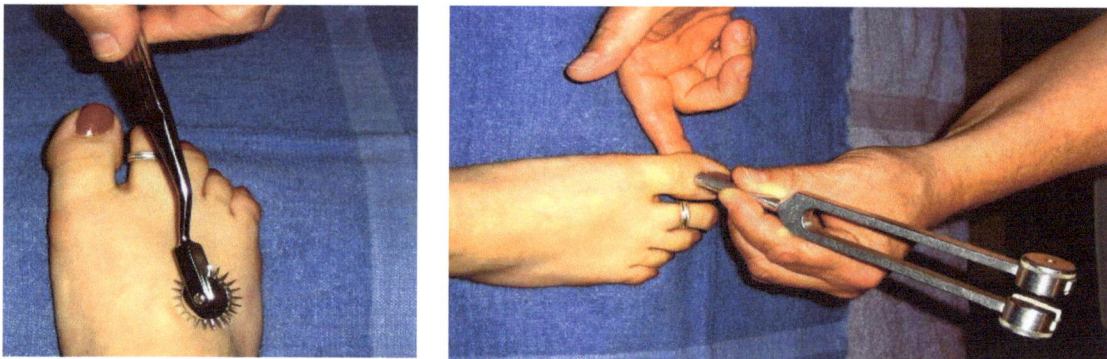

Figure 2, Physical Examination for Neuropathy. While there are many sophisticated diagnostic tests to document peripheral neuropathy, the diagnosis can usually be made with a thorough history and good physical examination. The image above left shows a pinwheel. The pinwheel will feel sharp in those patients without damage to their peripheral nerves. In neuropathy, the pinwheel may feel dull, may provoke numbness or tingling or may not be felt at all. The nerves that carry vibration sensations to the brain are often affected early in peripheral neuropathy. Many patients will not be able to feel vibration at all. They are usually surprised to find out that they can feel vibration on their ankle, at their knee or on their finger, but can not feel this sensation in their toes. Neuropathy patients will experience early decay in vibration sensations in their feet. The figure above right shows a tuning fork placed on the patient's great toe. The neurologist places his finger against the bottom of the patient's toe and can actually feel the vibration of the tuning folk through the patient's toe. In early decay, the patient will tell the doctor he can no longer feel the vibration. However, the doctor can continue to feel the vibration through the patient's tissues. In this way the doctor compares his normal appreciation of vibration sensation directly with the decreased appreciation of vibration sensation in very early nerve damage.

What causes Neuropathy?

Neuropathy is one of the most common conditions affecting the nervous system. The more common causes are complications from diabetes, alcoholism, exposure to toxic chemicals, chemotherapy treatments, prescription drugs, and certain metabolic disorders. Many times, the exact cause can not be identified; this is not uncommon. Patients without an identifiable reason for their neuropathy are classified as having idiopathic neuropathy. In other words, idiopathic neuropathy is neuropathy of unknown origin.

You will read many times in this book about the importance of finding the cause of neuropathy, when possible. Treating neuropathy is a difficult undertaking. It is even more difficult if the offending agent(s) are not eliminated. That is because nerves can break down and become damaged much more quickly than they can be repaired and regenerated.

You must do everything possible to reduce or eliminate new and ongoing nerve damage while you start the long and difficult process of repairing the existing nerve damage and reprogramming the signals traveling within the nervous system.

What exactly happens to the nerves?

There are several general types of damage that occur when a patient has neuropathy. One is that the insulation surrounding the nerve (known as myelin) becomes damaged. This is similar to an electrical wire that loses its insulation. The normal signals the nerve is designed to carry either can't get past the area were the myelin (insulation) is damaged, the signal may get scrambled; or when several nerves lose myelin, the signal may short circuit. Usually, a little bit of each of these problems occur together in any one patient. The result is abnormal sensations like numbness, tingling, burning, or other odd sensations. Depending on the type of nerve fiber affected and the severity of the damage, weakness and even muscle wasting can occur. Unfortunately for the neuropathy patient, these scrambled signals frequently transmit pain sensations, even when there is no painful event occurring. We don't know why, but it is very common for abnormal pain signals to be transmitted at night. Most patients suffering from neuropathy have a difficult time sleeping because, for whatever reason, their pain becomes much worse at night. The second type of damage is when the actual nerve fibers themselves, rather than the insulation that covers the nerve, become damaged. *Figure 3.* This is more serious and can lead to muscle weakness or even muscle wasting. If the nerve fibers that carry non-painful information die and only the nerve fibers that carry painful signals remain, painful neuropathy will result. This patient is in constant pain with very little numbness, tingling, or burning sensations. Most patients have a combination of insulation damage and actual nerve damage, so their symptoms will vary depending on what is happening in their peripheral nerves.

The damage to the insulation and nerve fibers cause many chemical changes within the nerves. The altered signals traveling to the brain which occur in damaged nerves, cause chemical and even structural changes in the brain itself. These changes in the nervous system are collectively called neuroplasticity and will be discussed in later sections.

The process of neuropathy is extremely complex, involving many pathways and chemical cascades in the body. Neuropathy may start as a localized condition of nerves in the hands or feet, for example, but will ultimately cause chemical and structural changes in the spinal cord and eventually the brain. [1] That is why attempts at treating neuropathy require multiple measures and multiple approaches. Complex problems require complex solutions. Neuropathy is a very complex problem.

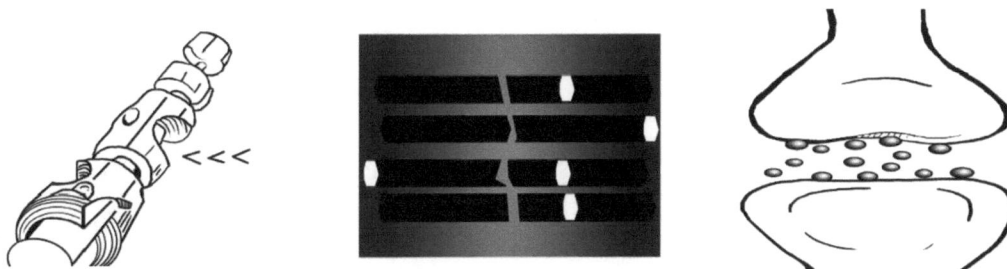

Figure 3: A, B & C, (A above left), is an illustration of demyelination of a nerve. This process is like when insulation is damaged around an electrical wire (see the <<<<). Short circuiting of nerve signals occur and normal function and communication along the nerve is lost. **B** The middle illustration above demonstrates denervation. This is when the nerve cell axons actually break. Using our example of an electrical wire, it would be like cutting the copper inside the wire with wire cutters. When the nerve axon breaks, this totally prevents necessary nerve signals from reaching their intended destination. Furthermore, denervation results in sprouting and the development of hypersensitivity discussed in the previous section. **C.** Above right is an illustration of a neuron synapse. Unlike electrical wires, nerves carry their messages not only through electrical impulses, but also by chemical messages. Demyelination and denervation can cause chemical abnormalities in the body's communication systems. In addition to the physical interruption in the transmission of important information, it is possible for chemical changes to persist even after the original nerve damage has been repaired. This process is known as pathological neuroplasticity. The point of these illustrations is that both physical and chemical disruptions occur in the nervous system of patients suffering from neuropathy. Usually all three of these pathological conditions occur in varying degrees in any one neuropathy patient. Treatment must address all of these problems.

What is Alternative Medicine?

There are several terms you need to be familiar with. One is *Alternative Medicine;* the others are *Complimentary Medicine* and *Integrative Medicine.* Alternative medicine involves those treatments, therapies, and remedies that are not routinely practiced by medical doctors or in hospitals - things like Chiropractic, Acupuncture, Ayurveda and Herbal Medicine. The name implies that these therapies are an "alternative" to the more traditional practices of medicine.

Complimentary Medicine consists of the same types of therapies utilized in alternative medicine, but not as an alternative to usual medical care, but as a compliment to medical care. Complimentary medicine is often care provided outside the mainstream. Too often, this care is provided without the knowledge of the patient's primary care provider or family doctors. Integrative Medicine also involves the use of alternative therapies. Not as an alternative to routine medical care or an ad-on compliment to medical care, but rather an integral part of medical care. Integrative Medicine seeks to "integrate" the best practices in medicine with the best alterative practices in a coordinated manner, creating a customized treatment plan for the patient based on which group of therapies will likely benefit the patient the most. The integration of these therapies is also done in such a way that will reduce the possibility of mixing therapies that may actually harm the patient. Integrative Medicine also attempts to insure that effective treatments prescribed by one practitioner are not inadvertently neutralized by therapies prescribed by other well-intentioned practitioners. The concepts are very similar, but integrative medicine seemingly provides the optimal situation for the patient.

What is Integrative Neurology?

First, we need to define the specialty of neurology. Neurology is the study of the nervous system and diseases of the nervous system.

Board Certification: To become board certified a practitioner must undertake a concentrated field of study (like neurology), then pass exhaustive tests known as board examinations and meet all the ongoing requirements for recertification each year. This process itself has nothing to with a doctor's license. It is a way for patients to identify doctors who have specialized education in a specific area.

Integrative Neurology is the integration of the medical specialty of neurology with non-traditional practices like chiropractic, acupuncture, herbal and other alternative practices. In a nutshell, it is combining specific knowledge of diseases of the nervous system, with treatments and therapies not routinely available to patients suffering from these neurological conditions.

The difference between a doctor's license, and his specialty

I am licensed by the States of Pennsylvania, Arizona and California through the State Board of Chiropractic. I am licensed as a doctor of chiropractic. Thus my credentials are "DC", Doctor of Chiropractic".

I am also board certified by the American Chiropractic Academy of Neurology, thus my credentials are DC, DACAN: Doctor of Chiropractic, Diplomat of the American Chiropractic Academy of Neurology.

Other types of neurologists are licensed under either the Medical Board, having a Medical Doctor's license, or MD, and specialized training in neurology, or they could be licensed under the Osteopathic Board and receive a DO, or Doctor of Osteopathy, license with certification in the study of neurology. The designations DC, MD, or DO are the doctor's licensing credentials. The board certification describes his or her area of post-graduate specialization and focused education.

Patients suffering from neuropathy should seek out a doctor with board certification in neurology because neuropathy is a disease of the nervous system. Most diseases of the nervous system are difficult to treat. Neuropathy is a complex and difficult condition. The field of integrative neurology offers some promising alternative treatments that are typically not available within the realm of the traditional management of neuropathy.

Treatment: There is hope

The first and foremost treatment in any case of peripheral neuropathy is to stop the ongoing damage to the nerves. This means the cause of the neuropathy must be identified wherever possible and the offending agent or agents reduced or eliminated. For example, treatment of diabetic neuropathy will require that blood glucose levels be tightly controlled. Utilizing the HbA1C blood test gives a good indication of the average blood sugar levels over time and also indicates how much sugar is binding to cells, such as neurons. Even though day-to-day blood glucose levels may be close to a normal range, the HbA1C test will tell how high the levels are in between blood tests. It is very difficult to treat diabetic neuropathy, if blood sugar levels are not well controlled or if tissues are binding sugar above 7%. As another example, relief of alcoholic neuropathy will be inadequate unless alcohol consumption is stopped. If toxins are the cause of the patient's neuropathy, ongoing exposure to the offending agents must be eliminated. It is fruitless to try to rebuild the damaged nerves and control or eliminate the signs and symptoms of neuropathy unless the cause of the nerve damage is eliminated. If an endocrine problem exists, the hormonal or other metabolic abnormality must be corrected. If medications are responsible for the patient's neuropathy, it is important to work with the prescribing physician to try and find alternative medications. Only after the offending condition that is provoking nerve damage has been eliminated or minimized can optimal nerve repair occur. In the case of chemotherapy-induced neuropathy or where the cause cannot be determined, reducing exposure to the offending agent is, of course, not practical or feasible.

Medical or Osteopathic Neurologists are probably better suited by training and licensure to provide the initial diagnostic work-up for patients suffering from neuropathy. They can order the necessary blood work and other studies needed to try to identify the underlying cause of a patient's neuropathy.

If a condition such as diabetes or a thyroid problem is identified, or a medication problem exists, they can prescribe or modify the necessary drugs needed to reduce or eliminate the underlying reason for the nerve damage in neuropathy. Treatment of the actual symptoms of peripheral neuropathy is another matter. Medical and Osteopathic Neurologists rely heavily on pharmaceutical management of the symptoms of neuropathy. At the time this book was written, there were very few pharmaceutical options for the symptomatic treatment of peripheral neuropathy, so there are very few medical options for treating the symptoms of neuropathy. There are virtually no traditional options with potential to repair and regenerate damaged nerves.

The following are the medications currently FDA approved as indicated for neuropathic pain. Duloxetine or Cymbalta[2, 3] is FDA approved for painful diabetic peripheral neuropathy. Pregabalin or Lyrica is FDA approved for post-herpetic neuralgia and painful diabetic peripheral neuropathy. [4]

At the time of this writing, these two drugs represented all the FDA approved pharmaceutical treatments for the symptomatic management of peripheral neuropathies.

It should be clear that while the medical neurologists are excellent in diagnosing and investigating the underlying causes of neuropathy, their ability to treat it is severely limited. They simply do not have many FDA approved options available for treating the pain of neuropathy. They have none FDA approved for repairing nerve damage in neuropathy.

Adverse Effects of Prescription Medications

To make matters worse, these approved drugs carry with them their own set of adverse and unwanted side effects. Below is an excerpt that lists some the unwanted side effects of the approved drugs for peripheral neuropathy.

Duloxetine or Cymbalta [5]

About one-quarter of patients enrolled in clinical trials stopped taking duloxetine after less than three months because of adverse effects. More than 40 different types of adverse effects have been reported, including suicide attempts and potentially severe hepatic disorders. Duloxetine is metabolized by the cytochrome P450 isoenzymes CYP 1A2 and CYP 2D6, creating a risk of interactions with other drugs that follow these metabolic pathways.

Diabetic Peripheral Neuropathic Pain
Approximately 14% of the 568 patients who received Cymbalta in the DPN placebo-controlled trials discontinued treatment due to an adverse event, compared with 7% of the 223 patients receiving placebo. Nausea (Cymbalta 3.5%, placebo 0.4%), dizziness (Cymbalta 1.6%, placebo 0.4%), somnolence (Cymbalta 1.6%, placebo 0%), and fatigue (Cymbalta 1.1%, placebo 0%) were the common adverse events reported as reasons for discontinuation and considered to be drug-related (i.e., discontinuation occurring in at least 1% of the Cymbalta-treated patients and at a rate of at least twice that of placebo).

Pregabalin or Lyrica

Controlled Studies with Neuropathic Pain Associated with Diabetic Peripheral Neuropathy

Adverse Reactions Leading to Discontinuation

In clinical trials in patients with neuropathic pain associated with diabetic peripheral neuropathy, 9% of patients treated with LYRICA and 4% of patients treated with placebo discontinued prematurely due to adverse reactions. In the LYRICA treatment group, the most common reasons for discontinuation due to adverse reactions were dizziness (3%) and somnolence (2%). In comparison, < 1% of placebo patients withdrew due to dizziness and somnolence. Other reasons for discontinuation from the trials, occurring with greater frequency in the LYRICA group than in the placebo group, were asthenia, confusion, and peripheral edema. Each of these events led to withdrawal in approximately 1% of patients.

These side effects are quite rare and if the patient experiences any of the these, immediate medical intervention should be sought. These include anxiety, confusion, incessant bleeding, chest pain, difficulty in breathing, loss of balance, muscle pain, swelling of lips or tongue, and rash. These are fairly common and require medical intervention only if they persist for a longer duration or are extremely severe in nature. Some of these include blurred vision, double vision, and loss of memory or tremors.

The side effects are very common and usually do not call for any medical intervention, unless they become extremely bothersome. Some examples include swelling in hands or ankles, constipation or diarrhea, dizziness, drowsiness, dry mouth, headache, insomnia and weight gain.

Considering the limited resources available to medical neurologist and the adverse effects that are associated with these standard medical options for the treatment of neuropathy, it should come as no surprise that patients are seeking alternative treatment options outside the medical mainstream.

Fortunately for suffering patients, many of the tools of alternative medicine can be integrated into the traditional medical management of peripheral neuropathy. This integrative neurology approach offers some very promising and hopeful treatments to patients suffering from peripheral neuropathy. Hope that is seldom found in traditional treatment approaches to of conditions of the nervous system like neuropathy.

The rest of this book will provide you with increasing detail of the pathology of peripheral neuropathy. Discussing the pathology of the condition will give you a better understanding of the complexity of peripheral neuropathy and will also provide insight into potential alternative treatment options.

Physical Treatments for Neuropathy

They are called physical treatments because they are mechanical hands-on treatments that, for the most part, do not involve drugs or chemicals. Let's look at how these treatments can be used to help patients suffering from neuropathy.

Electrical Nerve Stimulation

As we have previously discussed, nerves are like cables that carry hundreds, if not thousands, of individual wires. In the human nervous system, peripheral nerves are bundles of individual nerve cells called neurons. It is more technically correct to say that peripheral nerves carry hundreds, if not thousands, of *axons* which are a specialized part of the nerve cell.

If you examine peripheral nerves under a microscope, you will see that the axons that are bundled in a peripheral nerve are comprised of axons of differing sizes and shapes (see figure 4). Some are much smaller than others; some have varying degrees of insulation surrounding them. They range from no insulation (called myelin) to thickly covered (myelinated) nerve axons. Figure 4 illustrates the cut end of a peripheral nerve with the smaller axons highlighted in red and the larger myelinated nerve fibers highlighted in blue.

Generally speaking, the different sized nerve axons tend to carry different and specific information along them. For the most part, those axons that have little or no insulation (the smallest fibers in a nerve) carry painful sensations from the periphery to the brain for processing. So if you cut your finger, stub your toe, or you have nerve damage from neuropathy, it is more than likely that there is increased activity (in the form of pain signals) traveling from your body to your brain. The larger, well insulated nerve axons are, for the most part, responsible for carrying information about all other sensations (other than pain) to your brain.

If it is a painful sensation, it likely travels along a small nerve axon; if it is non-painful, like the wind blowing against your skin, or a butterfly crawling up your arm; it is likely carried by a larger nerve axon.

The last concept you'll need to know is that larger nerve axons carry information faster than smaller nerve axons. In essence, painful sensations and non-painful sensations "race" up toward your brain. The interesting thing about this race is that under normal circumstances the non-painful sensations get to the brain first, and they actually keep out or block the arrival of painful sensations (the ones that get there a little later). Think of this as a gate. The messages traveling along the large nerve axons get to the brain first and "close-the-gate" so painful sensations can't get into the brain. This concept is called the Pain-Gate Theory. Our example here is over simplified, but if you understand that some nerves "close-the-gate" on pain, you'll have a basic understanding of how virtually all techniques of electrical stimulation work to reduce pain.

Remember this about electrical nerve stimulation - stimulating the correct nerves, in the correct manner can "close-the-gate" on painful sensations.

Another property of electrical stimulation of nerves that makes these techniques so attractive for using them in painful conditions like neuropathy is the fact that electrical stimulation has a tendency to stimulate the large fibers more so than smaller fibers. This is the exact situation we want to activate the "pain gate" and block out pain messages from reaching the brain. [6]

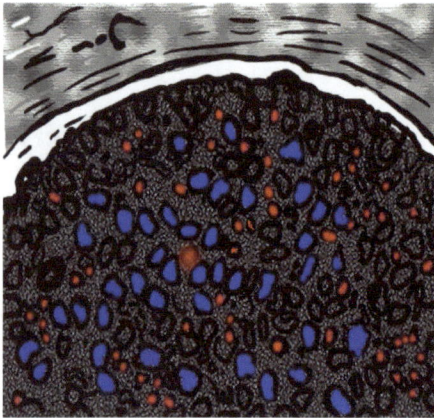

Figure 4, The illustration to the left represents the cut end of a peripheral nerve. It demonstrates small fibers which carry pain signals (red) and large fibers (blue) that carry non-painful sensations. These fibers form the basis of the pain-gate theory for using electrical stimulation to suppress pain. Since large fibers carry signals that tend to block pain and electrical stimulation preferentiality stimulates large fibers, electrical stimulation is an ideal therapy that can suppress pain.

These fundamental properties of electrical stimulation form the basis of most, if not all, of the techniques used to treat patients in pain. Stimulation can be applied over the skin above a nerve (so called TENs treatment), can to applied to acupuncture needles inserted into the skin (known as percutaneous electrical nerve stimulation or PENS), can be implanted in the spinal cord (a spinal cord stimulator), or even be implanted deep inside the brain itself.

This is known as deep brain stimulation. Regardless of where or how electrical stimulation is applied to the nervous system, the key concept to remember is that stimulating nerves that carry non-painful sensations will "close-the-gate" on the nerves that are carrying painful sensations. Electrical stimulation can be applied anywhere along the course of the nervous system. The newest techniques involve non-invasive direct brain stimulation for pain suppression. The technique will be discussed in detail later in this paper.

Non-Invasive Brain-Based Neuromodulation

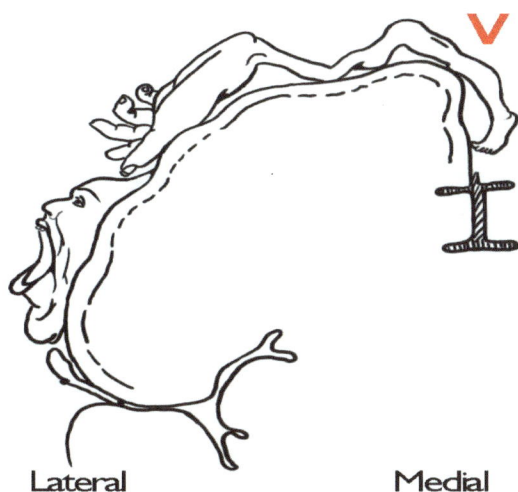

Figure 5, The entire body is represented on the surface of the brain. This representation is called a homunculus. Look closely at the illustration to the left. The part of the brain that represents the legs and feet is marked with a **V**. It is located near the very top of the head. This area is overly active in patients suffering from neuropathy and can be a target for direct brain stimulation. Stimulation using the negative electrode or cathode may suppress over activity in this part of the brain thus offering pain relief in the legs and feet. Targeting pain at the brain level makes direct brain stimulation for neuropathy pain relief a promising future therapy. It will be discussed in detail in later sections.

Lateral   Medial

Neuromodulation is the generic term now used to describe the techniques of pain suppression through electrical nerve stimulation. Before we get into the actual techniques of neuromodulation, it is necessary to talk just a little more about the anatomy and physiology of the nervous system. This discussion will as simplified as possible, but is necessary for a good understanding of how the various forms of neuromodulation work.

Before we start, now would be a good time to state that we really don't know exactly how neuromodulation works, but we do have a theoretical framework or clinical model of neuromodulation. As complex as the model seems, keep in mind that we have just begun to scratch the surface of our understanding of how the nervous system responds to electrical stimuli. Let's talk a little bit more about how nerves send signals to the brain.

At the extreme end of a neuron (nerve cell) are specialized endings called receptors. Receptors, for the most part, are like sensors that monitor the body and its environment and relay this information to the brain. We can use the senses as a good example to illustrate nerve receptors. The nerve endings in your ears are specialized receptors that sense sound. The receptors in the eyes respond to light. The receptors in the tongue respond to different tastes. In the skin and muscles, there are also receptors that respond to different types of stimulation. Receptors called *free nerve endings* are specialized pain receptors. All the other receptors, in general, are activated and respond to any and all sensations that are not painful. As we discussed in the previous section, the signals associated with pain are carried on smaller nerve fibers (the axons), and those messages that are non-painful travel on the larger axons.

There are exceptions to this general rule, but for our discussion of the complexities of neuromodulation, we will try to keep things as simple as possible.

It is important to have a basic understanding of receptors because they represent the first portion of the nervous system that can potentially be modified to help reduce the pain of neuropathy. If we can suppress the activity of *free nerve endings*, the receptors responsible for transmitting pain messages to the brain; we can provide some pain relief to patients suffering from neuropathy. How we might suppress the free nerve endings will be discussed in greater detail in the section on topical liniments later in this book.

If we trace a nerve fiber from its endpoint, the receptor, and follow it back towards the spinal cord, the next structure we can influence to try to reduce the symptoms of neuropathy is the axon. As we touched upon previously, the axon is very much like an insulated electrical wire.

Nerve cells and, in particular, nerve cell axons, carry electrical impulses, moving messages encoded within electrical signals from the receptors at the end of the nerve, along towards the spinal cord and then ultimately up to the brain.

You should remember from the previous section that larger nerve fibers that carry non-painful information compete with smaller fibers that carry pain-related signals. Anything we can do to stimulate the larger fibers will close-the-gate on the small fibers and their pain messages.

Transcutaneous electrical nerve stimulation (TENs) is a technique wherein electrodes are placed over the skin (see figure 10). A mild electrical current is passed through the electrodes. This current causes the nerve fibers under the skin to start to send signals. Luckily, the larger nerve fibers are more easily stimulated by this electrical current than the smaller nerve fibers. So you see, TENs therapy is a way for us to help the body close-the-gate on painful nerve signals like those seen in patients suffering from neuropathy. [7-12]

This nerve stimulation actually works with the body's own pain suppression system. The fact that larger nerve fibers are more easily stimulated by electricity than the smaller fibers is what makes pain suppression with electrical stimulation possible.

We still have a little more to discuss about nerves and their axons. We have used the analogy of an electrical wire to try to explain how nerve axons carry information from one part of the body to another. There is a difference however, between how electrical wires carry electricity and how our nerves carry electrical signals.

Our nerves are actually much more like batteries than true electrical cables. Just like in a battery, there are chemicals in our nerve cells. Just like we were able to use electricity to influence how nerves function, it may also be possible to further influence how nerves work by manipulating the chemistry of the nerves. To do this, we need another quick and over-simplified lesson in the microanatomy and the physiology of nerve cells.

Let's take an even closer look at the nerve axon to see if there is anything we might be able to manipulate to try to restore the nervous system to normal for patients suffering from peripheral neuropathy.

If you use a really strong electron microscope to examine a nerve axon, you will see holes on the surface. Theses holes are like pores in the surface of your skin. They function as a quick route to allow chemicals like calcium and sodium to enter or leave the inside of the nerve axon. [13-18]

We have talked about the pain gate theory previously, and these pores act like another gate. Controlling the chemical consistency inside the nerve and helping to keep it different from the chemistry outside the nerve cell. The important point to remember is that the quality of the chemistry inside versus outside the nerve is controlled for a large part by these pores. The pores act like channels.

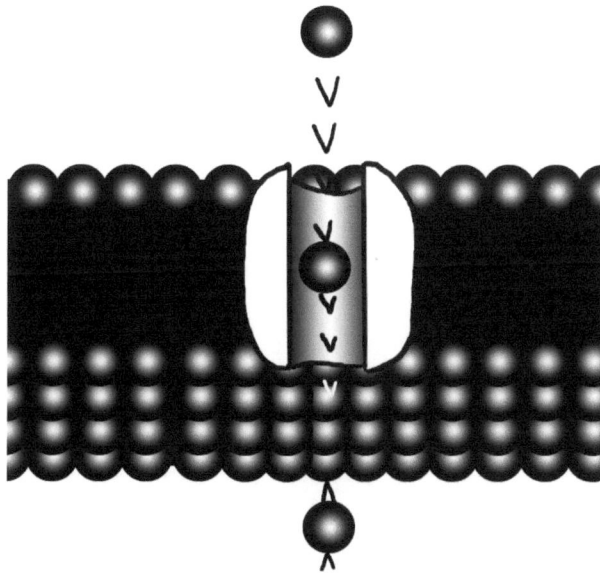

Figure 6. The illustration at the left demonstrates the channels found in the walls of the nerve cell axons. The purpose of these channels is to help maintain the chemical environment inside the nerve axon. The channels, by changing the chemistry of the axon, can render the nerve more irritable or less irritable. This means the signals traveling along the nerve can either be facilitated or suppressed. This has important implications for treating the irritated nerves of neuropathy.

In fact, the pores that are responsible for balancing calcium are called calcium channels, and the pores that are responsible for regulating sodium are called sodium channels.

The ease by which signals are carried along the nerve cell axon can be increased (the axon is more excitable) or decreased (the axon is suppressed and less excitable) depending on the state of these channels. This means that the channels in the nerve cell act as fine tuning controls and influence how signals are moved along from the body to the brain. In patients with neuropathy, the nerve cells tend to be more excitable.

Drugs that block calcium channels and sodium channels can suppress pain and other sensations by controlling the pores. You may have heard of the class of drugs called calcium channel blockers. These influence the nerves by closing the pores that regulate calcium in the nerve cell. New research has revealed highly specific types of calcium channels that are related to pain transmission. Sodium channels, when closed, will abolish signals traveling along a nerve axon. Both calcium channels and sodium channels are promising targets for suppressing pain associated with neuropathy. Figure 6 illustrates ion channels in the nerve axon. We will discuss them in more detail in the section on herbs and topical liniments that we use to help our patients suffering from neuropathy.

Moving along the path from the peripheral nerves towards the brain, we will need to discuss a structure call the *synapse*. A synapse is where one nerve ends and another nerve begins *(refer to figure 3C)*. So if we trace the pain in a patient suffering from neuropathy from her foot, the receptors (nerve endings) of those nerves will be found in and around the feet.

The axons of those nerves will run along the inside of the leg. The end of this nerve cell will be found at a synapse within the spinal cord. To refresh our understanding of pain signals, they start in free nerve endings in the feet, are carried toward the brain by small nerve axons. They then are passed off to other nerves at the synapse in the spinal cord. This passing off of nerve signals is accomplished by chemicals released in the synapse in the spinal cord. However, the synapse is not a simple link between one nerve and the next. There are many nerves that come together at the synapse. All release some type of chemical, creating a soup of different compounds called neurotransmitters. These neurotransmitters determine if the nerve signal stops at the synapse, if it is carried further along the nervous system towards the brain, how fast, how slow, how easily, and many other qualities of the nerve signal. At the synapse in the spinal cord, literally hundreds or thousands of inputs merge. The sum total of all these inputs determine what type of message, if any, will ultimately be transmitted up to the brain for further processing. To be technically correct, the synapse is where the pain-gate actually resides.

The synapse in the spinal cord is an excellent target to try to suppress pain signals ascending towards the brain. The synapse can be controlled and the final signal modified by changing the various chemicals released into the "soup" at the synapse. Two important chemicals elated to the treatment of neuropathy that reside at the synapse between nerves are GABA and Acetylcholine.[19, 20] We'll discuss them in detail later. The synapse can be influenced by input from unrelated nerves, through drugs, through herbs, electrical stimulation, acupuncture, spinal manipulation, and many other means.

Unfortunately, the synapse and the chemical soup found at the synapse can also be influenced in a negative way by abnormal inputs from damaged nerves like those found in patients suffering from neuropathy. You could say that winning the battle of the spinal cord synapse will determine how effective a treatment will or will not be for the neuropathy patient. Controlling as much as possible what goes on at the synapse is that important!

The nerve signal leaving the synapse and ascending toward the brain will be modified and altered at other relay stations and synapses in the higher levels of the spinal cord, brainstem, and brain proper.

The basic circuit we just described is more or less repeated as we follow a nerve signal up to the cerebral cortex, the highest part of the brain. As the nerve signal ascends there are additional opportunities to try to suppress it, or at least modify it, to make it seem less painful to the neuropathy patient.

As you can see, the complexity of the nervous system is a dual-edged sword. It makes things difficult for us to even understand, yet alone influence, but also provides us with many opportunities to shape and modify messages traveling within it. There is no question that as our understanding of the nervous system improves, so will our ability to safely and effectively influence it to the advantage of our patients. Even though our understanding of the nervous system is limited, we still have many avenues to pursue in helping our patients control the pain of their peripheral neuropathy. That is what this book is all about. That is why there is hope for neuropathy patients.

Electrical Stimulation

Now that we at least laid the groundwork for how and why neuromodulation may work, we can discuss studies and specific stimulation techniques and how they may help neuropathy patients.

Transcutaneous Electric Nerve Stimulation, or TENs, has a long history of use as a treatment to suppress chronic pain. The theory that explains the effectiveness of TENs was first published in 1965. Other emerging techniques utilizing electrical stimulation for pain suppression appear promising for patients suffering from peripheral neuropathy.

As our understanding of the nervous system's innate pain suppressing circuitry grows, so do reports in the scientific literature describing safe and simple stimulation techniques for pain suppression through neuromodulation. Also, newer applications of older neuromodulation techniques, like TENs, are being described for better and more effective neuromodulation of pain. The following section will discuss reports of promising new techniques for suppression of pain through electrical stimulation and modulation of innate pain suppressing circuits.

Among the more promising of these techniques, due to their safety, simplicity, and low cost, are transcutaneous electrical vagal nerve stimulation in the neck, vestibular stimulation at the ear, and direct current brain stimulation over the scalp.

At first glance, the patient suffering from symptoms in her feet or hands might wonder how stimulation of the neck, ear, or scalp could possibly offer relief from painful neuropathy. To understand how these techniques might work in neuropathy, you must first realize that these techniques are not used to reduce pain locally at the neck, ear, or scalp, but rather they are used to non-invasively activate or deactivate circuits within the spinal cord and/or brain. Even though they are applied to the surface of various parts of the body, they act like remote controls, modulating deep structures within the brain and central nervous system. The discovery of these external switches is very exciting news. Consider that stimulation applied to the skin near the ear, can influence a nerve that indirectly controls deep brain centers like the Thalamus. The Thalamus is located in the core of the brain. It has, as one of its functions, the control and regulation of pain signals arriving from the body (like from the feet and legs), and the continuation of these signals to higher centers of the brain (*see figure 7*). We have talked repeatedly about "closing-the-gate" on pain. You can think of the Thalamus as another one of the nervous system's gates. The Thalamus is one of the last gates that can modulate pain signals before they reach the highest centers of consciousness in your brain. After the pain signals leave the Thalamus and reach your cerebral cortex, you will consciously feel, recognize, and appreciate the pain in your feet (or elsewhere in your body). If the Thalamic pain gate is closed, you simply don't appreciate that your feet burn or hurt. We might call this final pain suppression circuit Thalamic Pain Gate.

Until the external switches to deep brain centers like the Thalamus were discovered, the only option for using the Thalamic Pain Gate to suppress pain was to surgically implant electrodes deep within the brain.

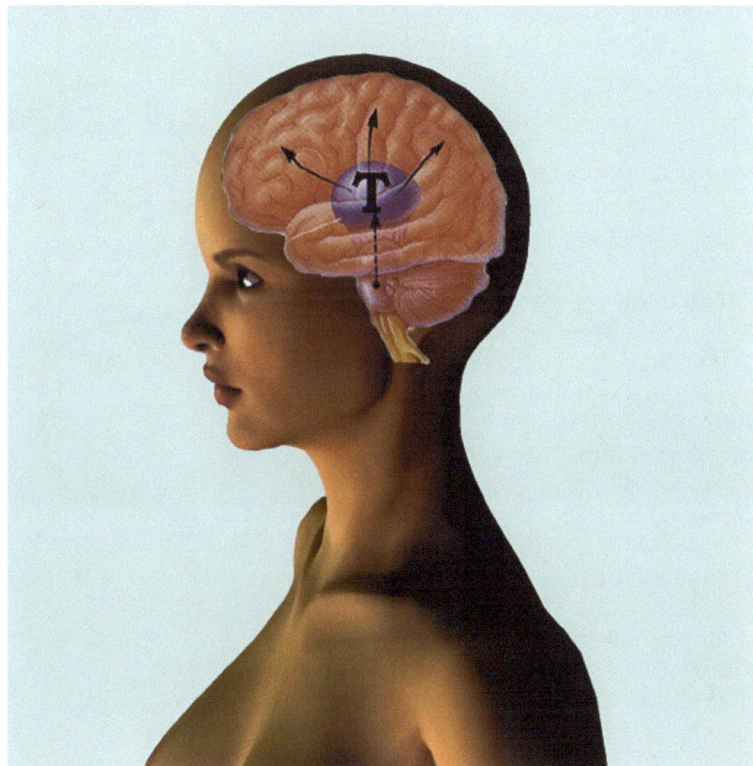

Figure 7. The Thalamus (T) is a structure located deep within the brain. It is a major relay station for pain sensations ascending from the body including the feet and legs. Non-invasive neuromodulation of the signals entering and leaving the Thalamus for pain control is a very promising technique with potential for helping patients suffering from peripheral neuropathy.

Consider the difference between having major brain surgery with electrodes and wires implanted within the brain versus a simple electrode placed on the skin around the ear. It will become obvious why transcutaneous activation of the Thalamic pain gate is so appealing to patients suffering from chronic pain.

What is interesting is that practitioners of acupuncture may have unknowingly been using these switches for hundreds of years. There is an entire system of acupuncture known as auricular acupuncture (ear acupuncture) that probably directly or indirectly modulates the Thalamic pain gate.

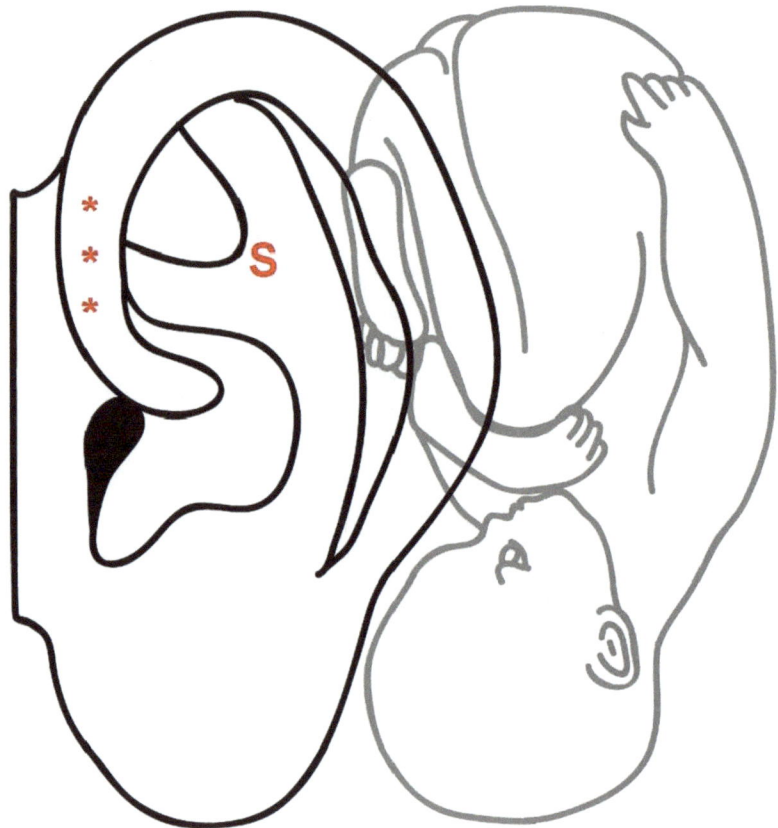

Figure 8. The Acupuncture Ear Chart to the right shows the point of stimulation for the feet marked with the red asterisks. The point marked "S" is known as Shenmen. Studies in animals suggest that stimulating these and other ear acupuncture points can actually change the chemistry of the brain. Electrical stimulation on the mastoid process just behind the ear will activate the vestibular system, which in turn modulates among other structures, the Thalamus. Thus the ear may serve as an external switch with which to influence deep brain structures like the Thalamus.

The classical acupuncture point known as Stomach-9, or ST-9, is also traditionally used to suppress pain. Interestingly, it is located in the neck very close to the site used to produce vagal nerve stimulation. The vagus nerve is one of the longest nerves in the body because it runs with the digestive tract. It also has projections that run into higher centers of the brain. Scientists discovered that electrical stimulation of the vagus nerve in the neck stops seizures.

Many patients who were receiving vagus nerve stimulation to control their seizures also had other conditions in addition to their seizures. Scientists, who were studying the effectiveness of vagal nerve stimulation in these seizure patients, began to look at how vagus nerve stimulation affected other symptoms in addition to its effects on their seizures. In many patients, vagus nerve stimulation not only reduced seizure activity, but also reduced painful symptoms like migraine headaches.

These observations led to experiments to see if vagus nerve stimulation could be used to treat pain. Early results have been promising. The technique for vagal nerve stimulation, at first, required electrodes to be implanted in the neck around the nerve. More recent studies are underway to determine if the vagus nerve can be stimulated through the skin, in much the same way as TENs is used. The electrical stimulation on the surface on the neck is very close to, if not exactly, where ancient Chinese acupuncturist stimulated ST-9 for pain relief. (*See figure 9*)

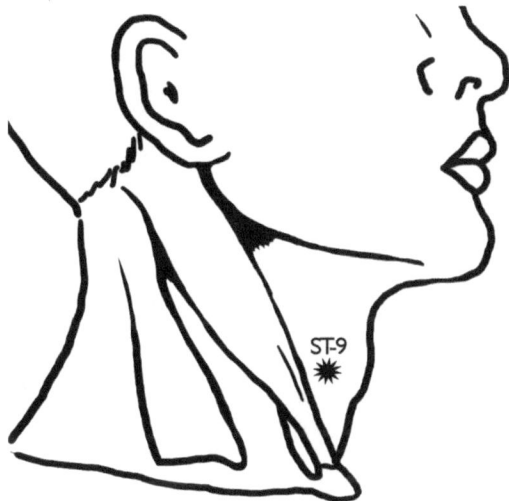

Figure 9: The acupuncture point known as stomach nine, or ST-9, is located near the carotid artery in the neck. The cervical portion of the vagus nerve is accessible at this location. This area is where vagus nerve stimulation is applied to reduce seizures. Stimulation of the vagus nerve at this site may also have potential to suppress chronic pain. The point was used by ancient Chinese acupuncturists to treat pain thousands of years ago.

This text has described the journey of pain signals in neuropathy patients that originate in the feet. It describes how these signals ascend from the nerve endings in the feet, through the peripheral nerve, into the spinal cord, ascending up to the brainstem and to the Thalamus. The final stop on this journey is the cerebral cortex. The cerebral cortex is the highest center of the brain. It is where we think and feel and understand our consciousness. It is the final opportunity to try to stop the pain signals which originated in the feet of a neuropathy patient. If there was some way to block the pain signal from our consciousness, the signal would still be there, the neuropathy would still be sending pain messages, but we simply wouldn't be conscious of them. In essence, we are ignoring these pain signals once they have arrived at the cerebral cortex.

Currently, the only way to block the conscious mind from recognizing and appreciating pain is to use powerful drugs. These drugs render the patient unconscious or heavily sedated. This is not a practical approach because it renders the neuropathy patient unable to function. Ideally, we would like to block the neuropathy patient's conscious perception of his or her pain without rendering the patient unconscious. Some of the newer techniques of neuromodulation of the pain gates may allow us to do just that. Let's take a look at some recent studies of neuromodulation.

TENs

As we discussed, stimulation of peripheral nerves with electricity (TENs) has been practiced for 40 or more years.[21] Recently, TENs has been studied specifically as a treatment for patients suffering from peripheral neuropathy.

Direct study of the effects of peripheral nerve stimulation in patients with Diabetic Peripheral Neuropathy has been carried out. It appears that stimulation improves nerve function in elderly patients with neuropathy. Treatment lasted 12 weeks. There was no reported improvement in the placebo control group of patients in this study.[22]

In another study of TENs in diabetic neuropathy patients, both lower extremities were treated for 30 minutes per day for three consecutive days. The patients' degree of symptoms and pain were graded daily on a scale of one to ten, before, during, and two days after treatment termination. Significant improvement in symptoms of neuropathy was reported. [23]

The symptomatic improvement following TENs treatment in neuropathy appears to be long lasting. In another study, done in animals, researchers report "unexpectedly long-lasting" pain alleviation from transcutaneous electrical nerve stimulation (TENs) in rats with peripheral neuropathy. [24]

For TENs groups, electrical stimulation for 16.7 minutes (1 Hz, paired current, 12 mA, 5-ms interval, 0.2-ms duration), once a day, was delivered for 5 consecutive days. Compared to the non-TENs groups, rats in the TENs groups showed significantly reduced thermal hyperalgesia (pain) for at least 3 to 7 days after treatment. These results indicate a possible long-lasting therapeutic effect of TENs. [24]

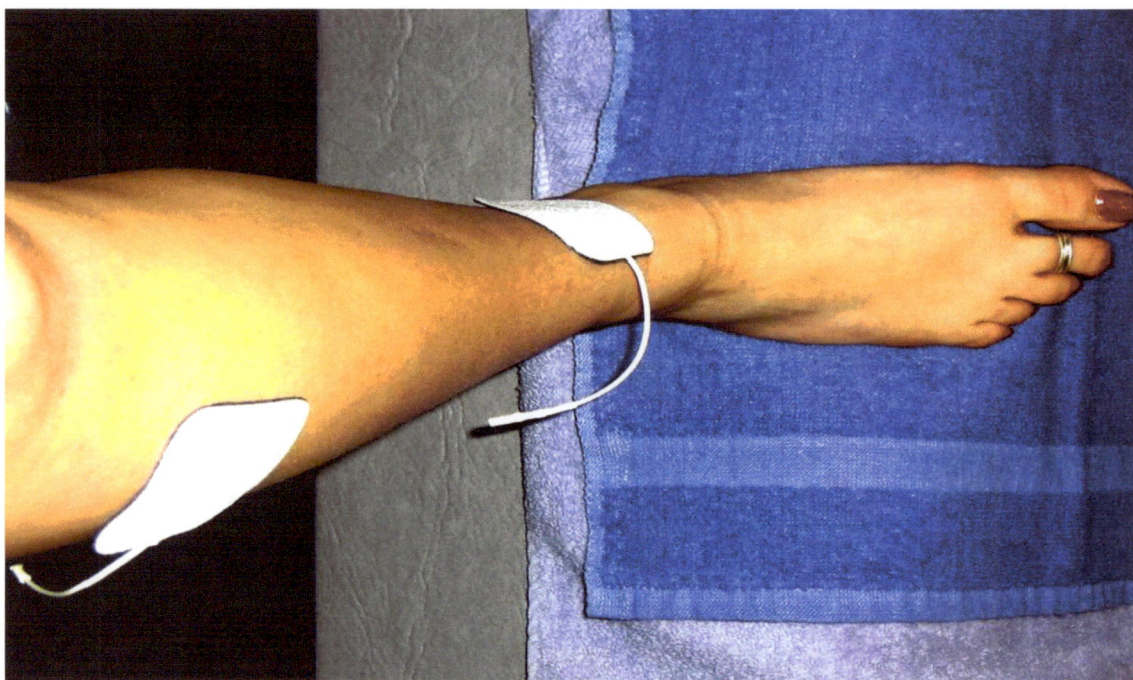

Figure 10, Placement of TENs electrodes over SP-6 at the medial ankle and ST-36 acupuncture point below and lateral to the knee are used in our office to relieve neuropathic pain.

Percutaneous electrical nerve stimulation or PENS is a variation of TENs and is also known as electro acupuncture. In this technique acupuncture needles are inserted into the skin and a electrical stimulation is applied through the needles (percutaneously).

The post-PENS treatment physical and mental domains of the SF-36 general health questionnaire and other outcomes tools all showed a significantly greater improvement with active PENs versus sham placebo treatments. Active PENS treatment improved the neuropathic pain symptoms in all patients.[11]

According to the results of this study, PENS may be a useful non-drug therapy for treating diabetic neuropathic pain. In addition to decreasing extremity pain, PENS therapy improved physical activity, sense of well-being, and quality of sleep while reducing the need for oral pain killing medications. [11]

In another study the overall improvement in neuropathic pain was also significant on an analog scale of 10 (p < .01), and correlated well with the percent amelioration data (r2 = .65). These data suggest an effectiveness of electrotherapy in managing neuropathic pain as an adjunct to the analgesics. Electrical stimulation appears to provide long term benefits with continued use, lasting in the responders group for over a year and a half. [25]

In animals the effects of TENs treatment appear to be associated with antinociceptive (pain reduction) as a result of a depressive action on the hyper-responsive spinal neurons found in this rat model of peripheral neuropathy[26]

This pain relieving effect of electrical stimulation and the changes in excitability of the spinal and other pain provoking neurons may be a result of modulation of powerful inhibitory neurotransmitters like GABA by TENs.

A study testing the hypothesis that either high or low frequency TENS applied to the inflamed knee joint would increase GABA in the spinal cord dorsal horn and activate GABA receptors spinally was conducted. Microdialysis was utilized to sample the extracellular fluid before, during and after TENS and analyzed GABA concentration in the extracellular fluid with high performance liquid chromatography. [27]

The researchers demonstrated that high frequency TENS increases extracellular GABA concentrations in the spinal cord in animals with and without joint inflammation. Similar increases in GABA did not occur in response to low frequency TENS. Furthermore there were no increases in glycine in response to low or high frequency TENS. Furthermore, the reduction in primary hyperalgesia by both high and low frequency TENS is prevented by spinal blockade of GABA(A) receptors with bicuculline. Thus, high frequency TENS increases release of GABA in the deep dorsal horn of the spinal cord, and both high and low frequency TENS reduce primary hyperalgesia by activation of GABA(A) receptors in the spinal cord.[27]

Electrical Stimulation Summary:

In summary, electrical stimulation either applied to the surface of the skin or closer to the actual nerve via electrical stimulation of inserted acupuncture needles, is believed to decrease pain and hyper-excitability of pain carrying neurons, through the elevation of the inhibitory neurotransmitter, GABA, in the spinal cord through selective and preferential stimulation of larger, more mylenated neurons in the peripheral part of the nervous system.

Vestibular Nerve Stimulation

The Vestibular System is that part of the ear and its central nervous system connections in the brain that deal with balance. [28, 29] The vestibular system is usually activated by motion or the perception of motion. Nerve signals from the vestibular system integrate with higher levels of the brain. [30-32] It is because of this connection, that some scientists have investigated how stimulation of the vestibular system may influence other areas of the brain, and if that influence can be used therapeutically. Experiments have determined the vestibular system in the ear has central connection to other brain regions like most notably the Thalamus and the Cerebral Cortex, [31, 33, 34] and including specifically the somatosensory and motor cortices. [31, 33, 34]

Stimulation of the vestibular system can be induced by spinning the patient in a chair, introducing cold water into the ear canal (the caloric test), by DC galvanic current applied to the mastoid process, or with High voltages Galvanic stimulation of the skin behind the ear. [35, 36] It is much more practical in a clinical setting to induce vestibular stimulation with either cold water at 0 degrees C for 15 seconds [36] [31] or through DC or AC Galvanic current at 1.0- 2.5 ma, 1-5HZ. [34] [36] Stimulation of the vestibular system to activate specific brain regions has a long and safe history in the research setting. [36] Only recently have these techniques begun to cross over into the clinical arena. [36]

Cold water vestibular stimulation has been shown in functional brain-imaging studies to result in activation in several contralateral cortical and sub cortical brain regions. [36] Using this method to activate these brain regions has been shown to have many positive therapeutic effects, including the suppression of chronic pain.[36] [37-40]. Other studies have demonstrated that vestibular stimulation can affect the motor cortex of the brain and produce measurable electromyographic signals recordable in the lower leg. [35] [41]. Because of the simplicity of the technique, its safety profile and relative ease of application, vestibular stimulation offers a promising new technique for neuropathy patients.

Vestibular Stimulation Summary: There are a number of non-invasive, relatively benign and safe methods for the stimulation of the vestibular system that show potential for treating patients with chronic pain syndromes. As the neural networks detailing the interactions of the vestibular and nociceptive (pain related) systems become better understood the potential for utilizing these methods to relieve the pain of neuropathy should grow.

Transcutaneous Vagal Nerve Stimulation

Vagus nerve stimulation in the neck has shown great promise in suppression of epilepsy. [42-47]. Because some patients who received vagus stimulation for their seizures also reported relief of pain, the technique has been studied in animal pain models [42, 46] and in experimentally induced pain in humans.[46,48]. Initial reports of pain suppression by use of vagal nerve stimulation has lead to it's application in chronic pain syndromes, like migraines and cluster headaches.[43,45]. Like vestibular and direct current brain stimulation, it appears that vagal nerve stimulation can be done non-invasively through stimulation of the vagus nerve transcutaneous. [43, 49-51]

It appears that vagal stimulation may initially reduce pain threshold with a long lasting rebound of anti-nociception occurring after stimulation is stopped.[42]

Vagal nerve stimulation appears to suppress activity in the spinothalamic tracts of the spinal cord and does not appear to be segmental in this inhibition. Indicating a central rather than peripheral pain suppressing circuitry is involved with vagal stimulation. [52-54]

The results of vagus nerve stimulation in human pain syndromes appear promising and may offer neuropathy patients another option to try to deal with their chronic pain. While there are currently no reports of clinical trials of vagus nerve stimulation specifically for neuropathy, an interesting observation related to polyneuropathy has been reported.

While the major pain generation in polyneuropathy is in the somatic peripheral nerves, pathologies in visceral nerves might be involved as well. Decreased vagal afferent activity is known to dis-inhibit pain perception, and therefore might contribute to pain in polyneuropathy. [55]

In experimental polyneuropathy created in animals with vincrisitne, a drug used to treat malignancy, those animals with intact vagus nerve function had much less signs of the symptoms of neuropathy than those animals who had their vagus nerves surgically severed. This data suggests that decreased vagal activity aggravates both the severity and the time course of painful polyneuropathy. Therefore, the two mechanisms: peripheral nerve damage and loss of vagal nerve function can add to each other in generating the pain picture of neuropathy.[55]

Summary: Vagal nerve stimulation may offer a non-invasive method for suppressing the pain of neuropathy.

Direct Current Brain Stimulation

Direct current is the type of current associated with a battery. It has a positive pole and negative pole. The anode is the electrode associated with positive polarity and the cathode is the negative electrode.[56, 57] [36] These electrodes have differing effects on biological tissue and recent research suggests that they can excite or suppress, respectively, activity in the nervous system.[56-61] Direct current stimulation has been used clinically for many years; however, only recently has it been studied for its ability to directly influence the function of the brain. What makes this modality so attractive and promising is that it can influence brain function transcutaneously, meaning it is a non-invasive method to change brain function from the outside in. [56-61]

Non-invasive direct current brain stimulation has been used to try to normalize brain function with promising results in stroke rehabilitation [62-65], Parkinson's disease [66-68], and some psychological and cognitive disorders [69, 70].

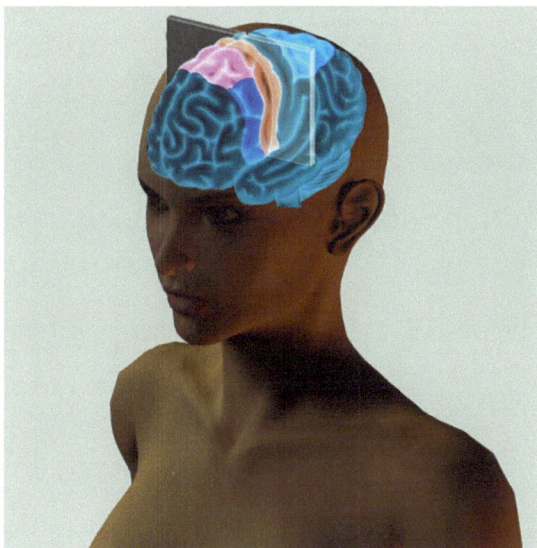

Figure 11: The location of the primary motor cortex known as M1 (pink). This is the site of electrode placement for non-invasive anodal direct current neuromodulation for pain suppression. M1 is located anterior to the central gyrus of the brain.

The area of stimulation for alternation of motor function is anodal stimulation over the primary motor cortex known as M1. The stimulating anode can be placed high on the apex of the skull to affect motor function in the legs *(see figure 5)* and more laterally over the upper extremity motor area for facilitation of arm motor function. The intensity of the stimulation is quite mild (1.0 ma) for the upper extremity, and more intensity at (2.0 ma) for stimulation of the leg area. The leg area resides deeper in the brain and thus further away from the surface electrodes. [56, 58, 71-73]

More recently, direct current brain stimulation has been applied to the suppression of chronic intractable pain states, including central pain syndromes and fibromyalgia. Emerging research suggests that direct current brain stimulation can provoke reorganization of neural circuits. [74-79] As was discussed earlier in this manuscript, neuropathy produces pathological neuroplasticity in both the spinal cord and the brain. Later in this section, we will discuss the potential of using direct current brain stimulation as a possible non-invasive method of correcting the pathological neuroplasticity of neuropathy.

While the exact mechanism of many brain-based therapies, including drug therapy, remains incompletely understood, it is believed that drugs that are effective in treating neurological conditions produce their clinical effects by changing the excitability of neurons in specific brain regions [80]. This theory of brain-based therapy provides the biological plausibility for the use of non-invasive electrical brain neuro-modulation.

As mentioned above, changes in cortical neuron activity in response to anode and cathode stimulation are very well documented. Experimental studies indicate that direct current stimulation can have long term effects after stimulation is discontinued. [80, 81] One reason for these long term effects seen may be due to reorganization and stimulation of neuroplasticity. [57, 82-85].

The effects of direct current brain stimulation on motor function have been demonstrated in stroke, Parkinson's disease, and other conditions of the motor system. [62, 63, 86]

Basic science studies of the brain suggest that final conscious appreciation of pain is associated with activation of the somatosensory cortex [81] Animal studies and more recently, human studies, have demonstrated that direct current stimulation of the motor cortex [81, 87] can have a reflex effect on the cortical excitability of the sensory cortex.

Motor cortex stimulation via the anode can reflexively suppress sensory cortex excitability and thus suppress pain appreciation [81, 87] Less studied is direct suppression of the sensory cortex through cathode stimulation of S1, the primary sensory cortex. [81, 87, 88]

Currently, various types of direct brain stimulation have been successfully used to treat a number of chronic painful conditions. Success has been reported in treating fibromyalgia and in correcting sleep disturbances associated with fibromyalgia. [73, 89, 90]

The suppression or facilitation of activity in higher pain processing regions of the brain form the basis for the use of direct current brain stimulation in the treatment of refractory pain syndromes.

Electrode placement: Optimal electrode placement appears to be over the primary motor cortex (M1) for the anode. By far, most studies have used this location for treatment. Cathode placement at S1 on the primary sensory cortex has been studied. Other locations and/or combination of anode and cathode placement should be studied to try to maximize the effectiveness of direct brain stimulation. Classical acupuncture points that are found on the scalp near the primary motor and sensory cortex respectively have been reported in ancient Chinese acupuncture texts to have pain suppression properties (*see figures 12-13*).

The relationship between classical acupuncture points on the Bladder and Gall Bladder Meridians to the motor and sensory cortices.

Acupuncture Points Gall Bladder Meridian

Classical Acupuncture points along the Gall Bladder Meridian course over the skull and among other brain structures, the primary motor and sensory cortex, respectively (see figure 12). [91] Specifically, the 15th numbered point of the Gall Bladder Meridian (described below) would be a good choice for stimulation of the primary motor cortex.

Chinese Name: GB-15 Toulinqi (English translation: Head Overlooking Tears)

Interestingly, this point historically was used to treat headaches and convulsions, so it has a long history that supports its choice for non-invasive brain neuromodulation of pain states such as peripheral neuropathy. The needle would be further stimulated with high voltage galvanic stimulation using the anode.

Historical & Classical Indications:

- Headache, nasal congestion
- Dizziness, lacrimation
- Infantile convulsion

Functions: Dispels wind; benefits the eyes, nose and head; calms the Shen; relieves pain.

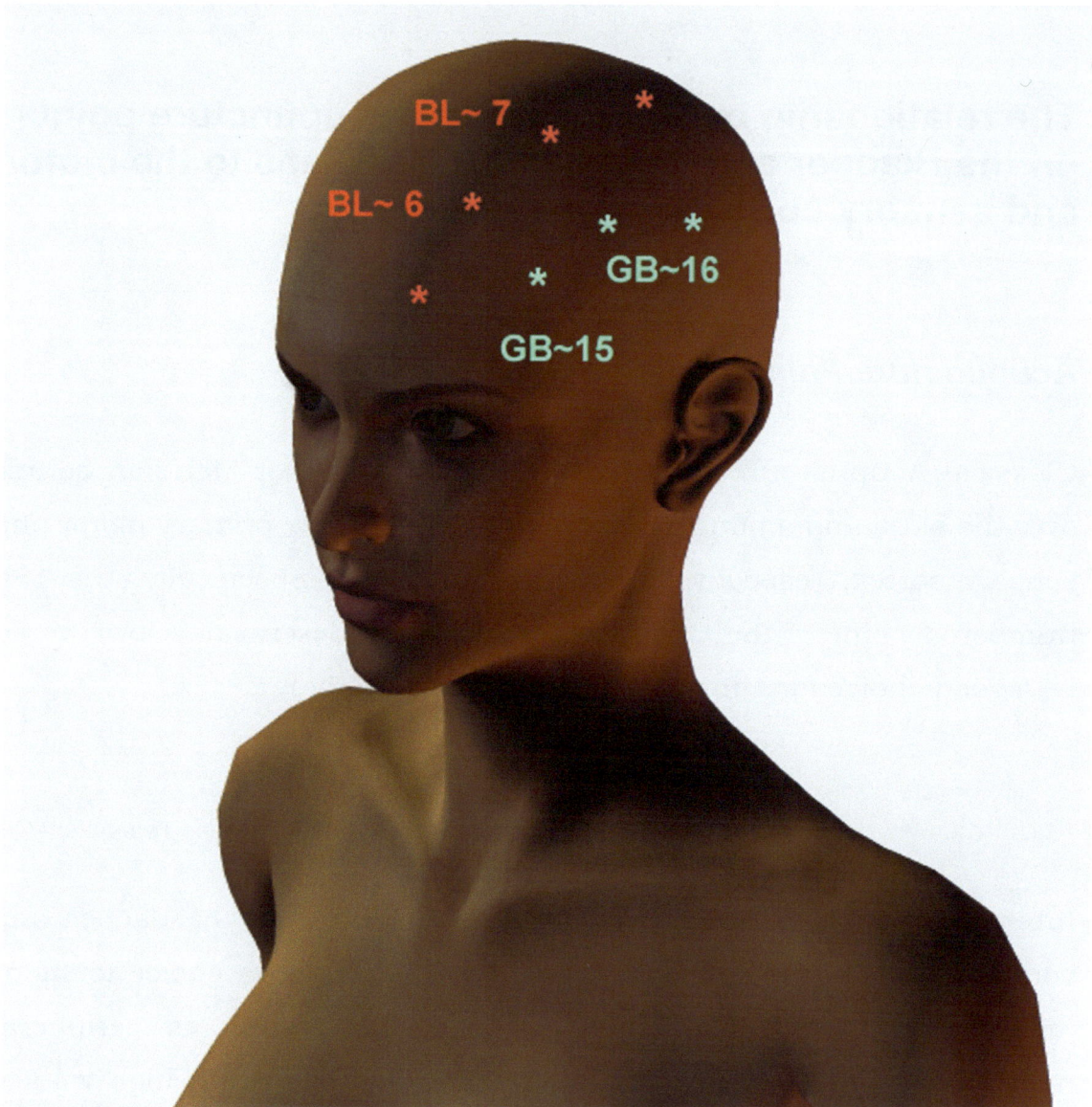

Figure 12: The Gall Bladder (GB) and Urinary Bladder (BL) Meridians in traditional Chinese acupuncture course over the head and scalp.

Acupuncture Points Urinary Bladder Meridian

Like the Gall Bladder Meridian, the Bladder Meridian also courses the skull. It is situated more laterally than the Gall Bladder Meridian and specific points of this channel may line up over motor and sensory cortical regions. [91] See the illustration of the homunculus in figure 5. The modern techniques of non-invasive neuromodulation use sponge electrodes which cover larger areas of the motor cortex. At this time, it is unknown if highly specific cortical anode stimulation using acupuncture needles at specific points or generalized cortical stimulation using sponge electrodes is more effective clinically. It is possible that both are equally effective. Specifically, Bladder Point Number 6 (BU-6 or BL-6) appears to be the optimal choice of a Bladder Point for stimulation of the motor cortex. Theoretically, one might insert needles on these Gall Bladder and Urinary Bladder Meridian Points associated with the primary motor cortex (connecting them to the anode for facilitation and stimulation) while simultaneously choosing other points in these meridians that are corresponding to the primary sensory cortex and using cathodal stimulation to suppress the activity at the primary sensory cortex. The best electrode placement and stimulation parameters are still being investigated. For clinical applications, the primary motor cortex stimulated with the anode has the most abundant research to support its effectiveness. A look at the historical use of Bladder 6 (Bl-6) is interesting. It is described below and has been used for centuries to treat headaches and paralysis.

Chinese Name: UB-6 Wuchu (English translation: Fifth Place)
Indications:
- Headache, dizziness
- Hemiplegia
- Epilepsy

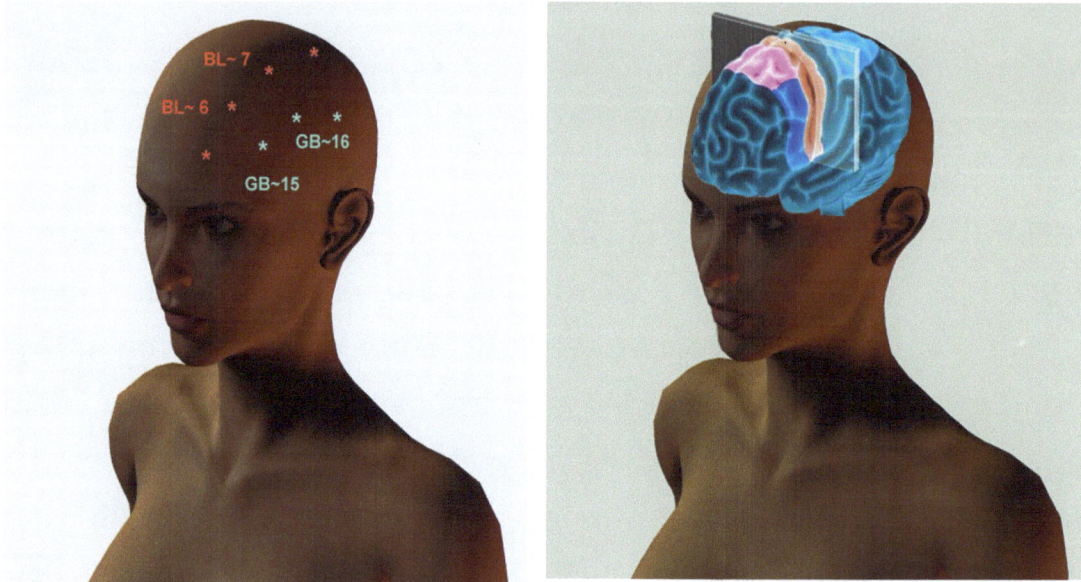

Figure 13. The relationship of Gall Bladder & Bladder acupuncture points to the underlying brain structures. Both of these meridians have points that overlie the brain including the primary motor cortex (M1) and Primary Sensory Cortex (S1) as seen above.

Stimulation parameters

Stimulation intensities ranging between 1.0 and 2.0 milli-amperes are the most commonly reported in studies of non-invasive direct brain stimulation. This is using a sponge electrode in most cases and low voltage direct current.

Warning

Low voltage direct current can cause chemical burns. The intensity of the stimulation will be concentrated and the potential for injury dramatically increased if low voltage current is applied through acupuncture needles.

High voltage galvanic stimulation carries much less risk of burning the patient; however, it has not been well studied, and it is not known if it has similar effects as low voltage galvanic stimulation on the nervous system. It is our unverified observation that high voltage galvanic is every bit as effective as direct current without the risks of burns.

These caveats must be considered and close patient monitoring is mandatory when using these techniques in the clinic.

Treatment Course

Long term alternation in brain pain perception appears to occur for stimulation that last in the neighborhood of 20 minutes and is applied for five consecutive days.

Safety Issues and Adverse Events

The safety of non-invasive direct current brain stimulation is well established. Adverse events reported in controlled trials were similar to placebo treatment and include dizziness, itching and burning under the electrode, and headaches.[80, 85, 92]

While not reported, seizures would appear to be a possibility in susceptible individuals. No cognitive deficits from non-invasive brain stimulation have been reported in controlled trials.

Informed consent and close patient monitoring is mandatory.

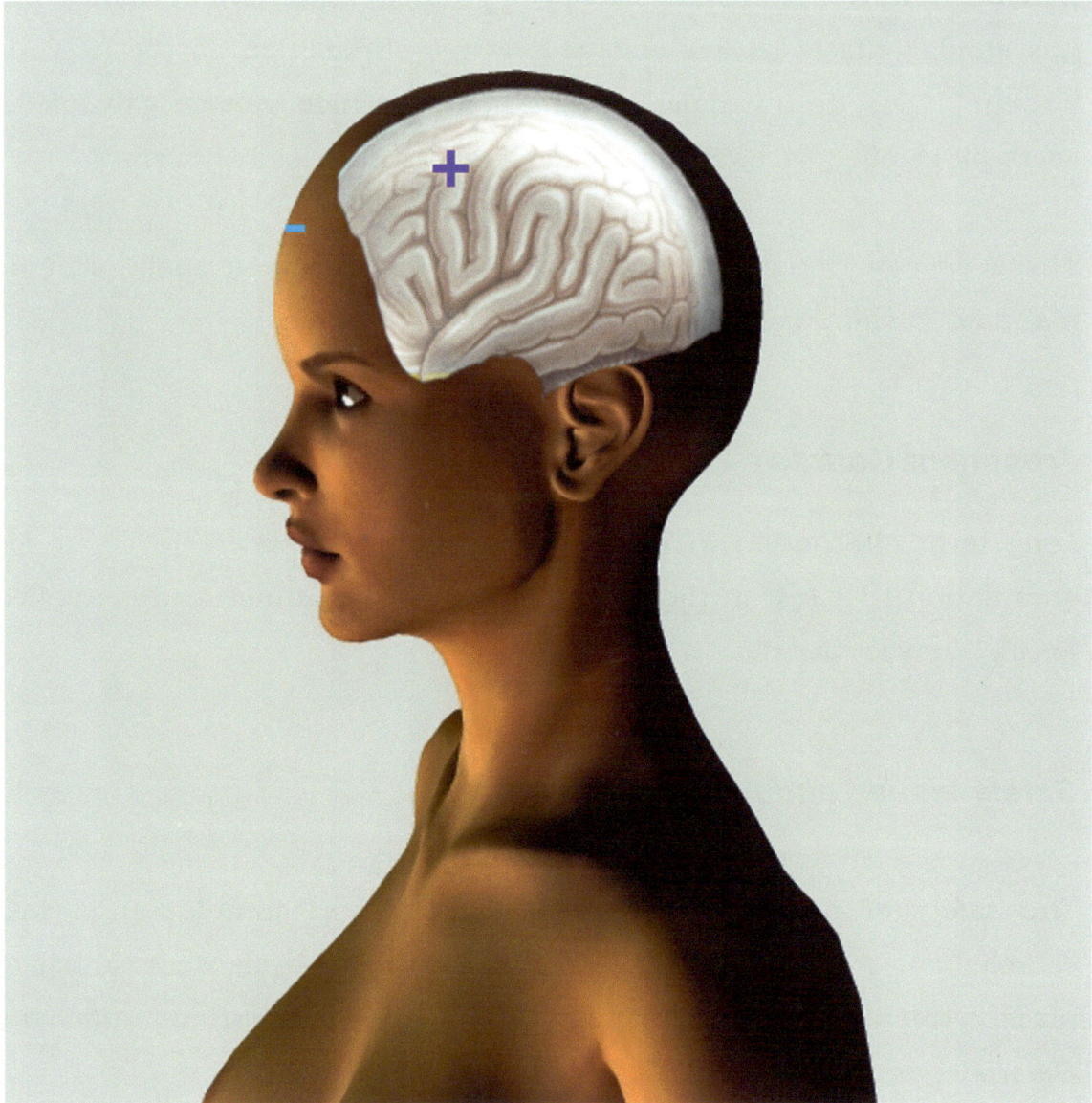

Figure 14 The positive (anode) and negative (cathode) electrodes are placed on the scalp over specific areas of the brain for the purpose of stimulating or facilitating function in these areas. The anode causes brain facilitation. Or if an area of the brain is overactive, the negative electrode (Cathode) can be used to transcutaneously suppress brain activity in the hyperactive region.

Laser Treatment of Neuropathy

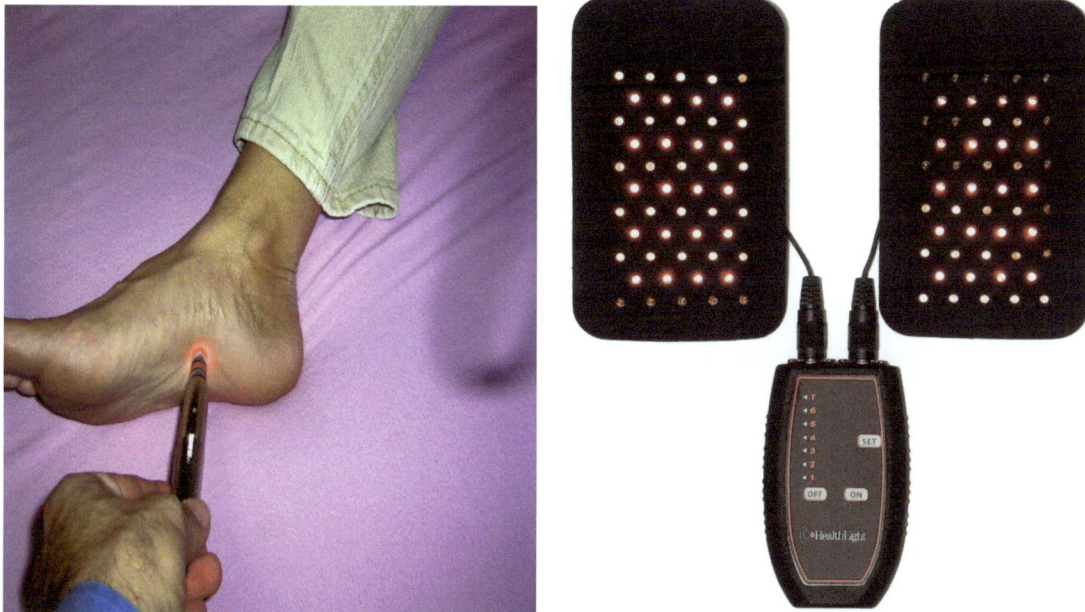

Figure 15, Low level laser therapy, above left (Laser-Therapy http://coldlasertherapy.us) hand held laser and multiple diode laser pads (Healthlight http://www.infraredtherapy.com) appear to have the ability to accelerate wound and tissue healing. These lasers are an integral part of the package of therapy we use to treat our patients suffering from peripheral neuropathy. Low level laser therapy is painless, safe, and does not even require that the laser touch the patient. The biological effects of this treatment make it a very promising treatment for patients suffering from peripheral neuropathy.

When we think of Laser Therapy, we tend to think of science fiction movies that show lasers cutting through steel plates or shooting down spaceships. The medical use of lasers and in particular, low power or low level lasers is growing dramatically.

Low power lasers don't damage tissues, they actually stimulate cell metabolism. There is a growing body of research that suggests that this low power laser treatment can accelerate wound healing, reduce pain and even promote nerve regeneration [93-103] We have had good clinical success in using our low level laser therapy to help diabetic skin ulcers in our neuropathy patients.

Laser therapy has several properties that seem ideal for the treatment of patients suffering from peripheral neuropathy. In patients with severe nerve sensitivity, a laser can be used and does not require any physical contact between the LED laser wand and the skin of the patient. The Laser light is beamed onto the skin, much like shining a flash light. It can be held stationary or moved in a "painting motion" see figure #15.

The two most common wavelengths for low level laser applications are 635nm and 880 nm. Some units use multiple LED wands that provide several different wavelengths simultaneously. The treatment is completely painless and because there is no contact between the laser and the patient, many patients doubt low level laser therapy can be a viable therapy for severe conditions like neuropathy. However, there is substantial research demonstrating the effectiveness of low level laser therapy. The volume of research supporting the use of low level laser therapy published in the scientific literature is increasing rapidly.

Low level laser has documented wound healing properties. In our office, we have seen first hand the remarkable ability of low level laser to heal diabetic ulcers, skin rashes, cuts, burns, and other conditions of the skin.

Research done in conjunction with NASA suggested that low level laser enhances natural wound healing and will more quickly return patients to pre-injury/illness status.[102]

There are a number of studies reporting wound healing effects as well as pain suppressing effects of low level laser therapy.[98, 103-106]

The healing effects of low level laser therapy are not limited to the skin and surface of the body, however. A number of studies document the effects of transcutaneous laser therapy (laser light shined over the intact skin) on tissue deeper in the body. Laser appears to have a potent effect on nerves lying under the skin.

Studies have shown that damaged nerves heal and regenerate better and faster when the skin over those nerves is treated with low level laser light. [93, 95, 96, 99-101, 107, 108]

The density of regenerated fibers was superior in the laser treated group compared with controls [95]. Laser light in the 633nm wavelength range up-regulated mRNA and increased the rate of nerve regeneration, successful reconnection, and nerve cell survival [100]. The microscopic appearance of inured nerves after treatment with low level laser therapy recovered to near normal after a course of laser therapy [93]. Laser treatment was found to produce significant amounts of positive structural and cellular change in the damaged ends of nerves [96].

In sciatic nerve injuries, laser treatment was associated with more reconnection of peripheral nerves and better quality of myelin than the control nerves, with greater functional restoration.[108].

The effect of laser stimulation on nerves is not limited to the nerves directly under the skin being treated. The positive regeneration and restoration occurs locally and also has the ability to stimulate repair in the associated spinal cord area. Leading at least one researcher to speculate that laser therapy may have an important role in spinal cord injury treatment and rehabilitation [107].

Laser light therapy has an immediate impact on the conduction of nerves under the area being treated [101]. Laser appears to be able to restore normal conduction as measured by nerve conduction velocity studies.[94, 97, 101, 103].

All of these properties and the specific effects of low level laser therapy on nervous tissue, make it an excellent therapy for the patient suffering from neuropathy. We have seen remarkable improvement in our neuropathy patients after several weeks of low level laser therapy combined with other therapy in our office.

Recently, the direct effect of low level laser therapy on diabetic complications including ulcer and neuropathy has been reported.[97, 103]. Overall, low level laser therapy appears to be effective in reducing pain and promoting tissue and nerve healing and regeneration both locally and at a distance. Initial studies of laser for the treatment of neuropathy are encouraging.

The therapeutic qualities of low level laser therapy coupled with the fact that it produces few, if any, unwanted side effects, suggest that laser therapy should be included in any comprehensive approach to the treatment of peripheral neuropathy.

Ultrasound

Ultrasound is a high frequency sound wave that can penetrate through the skin and muscles down to reach the underlying nerves. It has the ability to penetrate the body deeper than any other physical treatment. Ultrasound has several properties and characteristics that make it attractive for use in treating painful peripheral neuropathy.

First and foremost, ultrasound therapy can directly influence the nerves. Depending on the intensity and frequency of the stimulation, ultrasound can either increase or decrease the conduction of signals traveling along peripheral nerves [109].

Therefore, if applied properly, it can reduce the signals traveling along smaller nerve fibers, thus reducing painful signals reaching the brain. It can also increase the signals traveling in larger diameter nerve fibers, thus enhancing the pain gating mechanisms in the spinal cord [110].

Ultrasound has direct pain reliving characteristics [111].

Recent research indicates that the application of ultrasound can modify neuroplasticity at the spinal cord level [110, 111] . This is a particularly important quality when considering the neuroplasticity changes that occur in cases of peripheral neuropathy. At least in the animal model of diabetic neuropathy, adverse neuroplasticity changes are known to occur at both the spinal cord and higher brain levels. *п, ¥:

*II Spread of excitation across modality borders in the spinal dorsal horn of the neuropathic rat

Thus, following peripheral nerve injury, excitation generated in dorsal horn areas which process non-nociceptive information can invade superficial dorsal horn areas which normally receive nociceptive input. This may be a spinal mechanism of touch-evoked pain. [112]

¥: Functional imaging of allodynia in complex regional pain syndrome.

Non-painful stimulation on the non-affected hand activated contralateral primary somatosensory cortex (S1), bilateral insula, and secondary somatosensory cortices (S2). In contrast, allodynia led to widespread cerebral activations, including contralateral S1 and motor cortex (M1), parietal association cortices (PA), bilateral S2, insula, frontal cortices, and both anterior and posterior parts of the cingulate cortex (aACC and pACC). Deactivations were detected in the visual, vestibular, and temporal cortices. When rating-weighted predictors were implemented, only few activations remained (S1/PA cortex, bilateral S2/insular cortices, pACC). [113]

These changes in the nervous system appear to be associated with defects in the pain control (gating) mechanisms of the spinal cord.

Ultrasound applied peripherally appears to be capable of correcting some of the neuroplasticity associated with neuropathic pain [110].

These studies suggest that ultrasound is an ideal therapy to provide symptom relief for patients suffering from neuropathy. Even more promising are the studies that suggest ultrasound can stimulate and accelerate nerve repair and regeneration [114-117]. The stimulation parameters for nerve regeneration were 12 ultrasonic treatment sessions over 2 weeks.

Ultrasound was applied at a frequency of 1 MHz and an intensity of 0.2 W/cm2 spatial average temporal peak (SATP) for 5 min/day. After ultrasound treatments, there were larger numbers or more fully developed regenerated neurons in the ultrasound treated group, than the control group. Combining ultrasound with nutritional substances (discussed later in the paper) that may initiate nerve recovery and repair is a promising combination in the comprehensive management of peripheral neuropathy.[116, 118]

Most patients treated in our offices feel that ultrasound is a pleasant and soothing therapy which provides immediate relief of their neuropathic pain.

Acupuncture Treatment

Neuroanatomical versus traditional acupuncture

Traditional acupuncture is 2,000 to 3,000 years old. The original premise in acupuncture was that energy, called Qi (or Chi), flowed through the body in channels called meridians. The meridians were believed to carry energy in much the same way as arteries and veins carry blood. According to traditional Chinese Medical Theory, interruption of this normal flow of energy throughout the body would cause Qi to either accumulate in tissues or be unable to reach tissues. In either event, too much Qi or too little Qi, in tissues would cause dysfunction and eventually disease. Inserting acupuncture needles into the body was said to restore the normal flow of Qi throughout the body and allow injured or sick tissues to regenerate and heal. When and where acupuncture needles were placed to restore the flow of Qi was based on complex diagnostic formulas. These formulas were constructed based on analyzing patterns of signs and symptoms in a particular patient.

According to Traditional Chinese Medical practice, several patients with the same condition might demonstrate different patterns and thus be treated with different acupuncture prescriptions. In other words, patients suffering from neuropathy might have differing patterns. Some may have more numbness than pain. Others might have muscle wasting, while still the next neuropathy patient may have burning pain at night.

So according to TCM theory, even though each patient was suffering from neuropathy, they all will be treated differently according to the patterns they exhibit. Individual treatments would be based on this pattern analysis for the specific patient.

It is well beyond the scope of this text to delve too deeply into the complex sets of relationships that make up the Traditional Chinese Medicine pattern analysis matrix. However, in general, patients with long standing chronic pain, like patients suffering from neuropathy, fall into several discrete patterns within Traditional Chinese Medicine.

Painful conditions like those seen in patients with neuropathy generally are considered in association with Wind, Heat, Blood Stasis, Dampness, and Phlegm accumulation in the affected tissues. Depending on which of these patterns is present in a particular neuropathy patient, acupuncture points would be selected, and supportive herbs would be administered to "Dispel Wind", "Quicken and Move Blood", Drain Heat and Dampness" and "Clear Phlegm". [91]

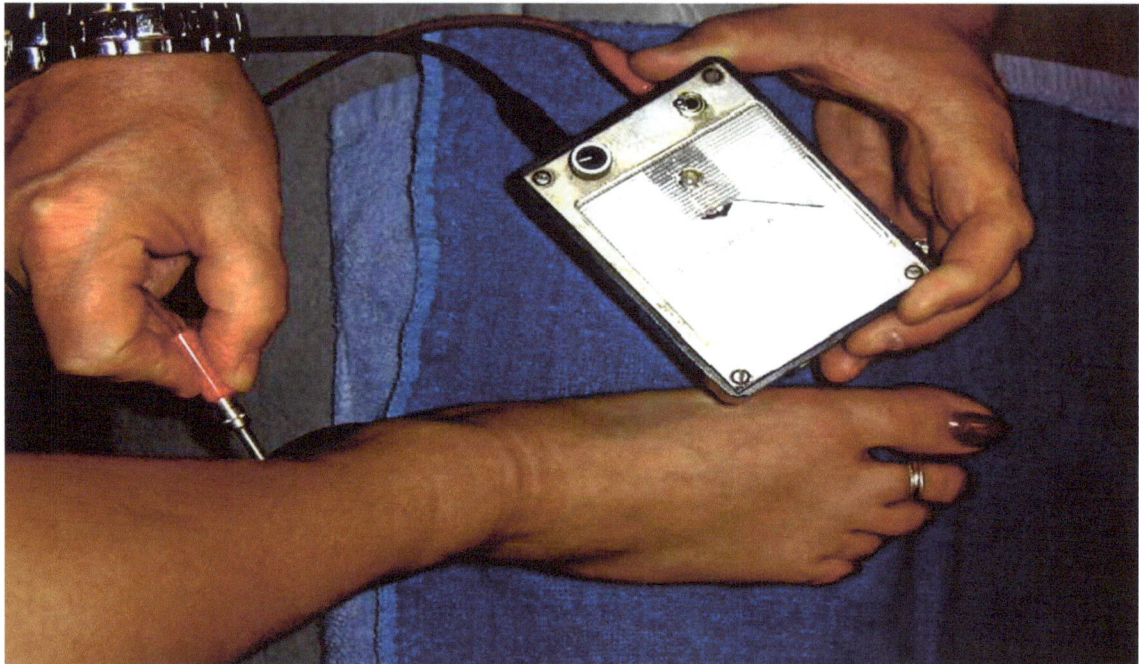

Figure 16 Acupuncture points have decreased electrical resistance which can be measured with a meter. We commonly use Spleen ~ 6 (SP-6) just superior to the medial malleolus for electrical stimulation and or acupuncture needling in our neuropathy patients.

Thus, the points would be chosen based on texts written thousands of years ago that listed specific prescriptions for treatments intended to "Dispel Wind or Clear Phlegm".[91] Utilizing traditional Chinese medical pattern analysis and treatment requires a leap-of-faith. These patterns are not recognized in Western Medical Science. Since the patterns are based on mysticism and folklore, the prescriptions for treatment based on these patterns are also based on superstition and folklore. There simply is no Western condition that is the same or even similar to the Chinese concepts of Wind or Dampness. It should be apparent that Traditional Chinese Medical Theory does not blend well with Western Scientific approach to patient care. However, Western trained physicians are turning to a science based model of acupuncture termed neuroanatomical acupuncture.

In neuroanatomical acupuncture, Meridians, Qi, and pattern analysis are replaced with peripheral nerves, nerve conduction studies, neuropathology, neurophysiology, and functional MRI imaging.

Since neuropathy is, by definition, a condition of the peripheral nerves, with pathological changes in the nervous system that are measurable using sophisticated neurological tests like electromyography and nerve condition studies, neuroanatomical acupuncture appears to be the ideal method with which to diagnose, treat, and hopefully restore nerve function in patients suffering from peripheral neuropathy.

Acupuncture For Neuropathy

Acupuncture in various forms has been studied for the treatment of neuropathy.

In Diabetic Neuropathy, Japanese and Chinese styles of acupuncture were compared. Both types of acupuncture were found to reduce patient-rated pain as measured on pain scales. Objective improvement in defective sensory perception was also demonstrated in the group receiving acupuncture using the Traditional Chinese Medical Model [119].

In another year-long study, acupuncture treatment of neuropathy was compared with best practices for medical management of neuropathy.

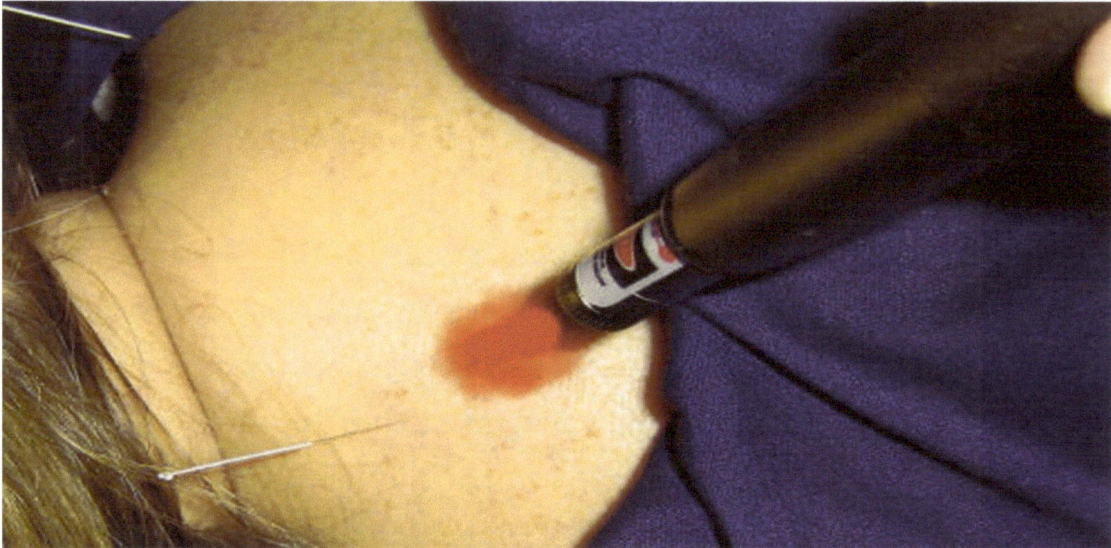

Figure 17, Combining traditional needle acupuncture with low level laser therapy in a patient with neuropathy. We call this method **Red Needle Acupuncture.** It combines one of the oldest therapies for pain relief, needle acupuncture, with one of the newest, low level laser therapy. This combination can be a powerful and effective treatment for patients suffering from peripheral neuropathy.

In the patients receiving the standard medical care (pharmaceutical drugs and self-help advice) only 15% of the patients improved. In comparison to usual medical treatment, 76% of the patients receiving acupuncture improved. Furthermore, the subjective improvements in pain reported by the patients receiving acupuncture correlated with objective improvement in nerve functioning measured by nerve conduct studies. [120]

These results suggest that acupuncture can be more effective than standard medical care for the treatment of neuropathy, and that acupuncture not only reduced pain, but also appeared to improve the functioning of the peripheral nerves. Most of the patients seen at our office for their neuropathy came to us for evaluation after trying the best standard medical treatment without benefit.

Our experience with treating neuropathy patients suggests, as this formal study has, that medical care for patients with neuropathy is usually inadequate.

Another study found that using electrical stimulation over specific acupuncture points for 30 minutes a day for 4 weeks in the animal model of diabetic neuropathy produced clinical improvement and functional recovery in nerves [121].

Using devices like TENS units and other forms of electrical stimulation was discussed previously in this text. This study suggests that it may be beneficial for neuropathy patients to not only receive electrical neuromodulation therapy, but to apply this stimulation over known acupuncture points.

The importance of proper placement of electrodes and stimulation over acupuncture points for the treatment of neuropathy pain, seems to be supported by other studies.

Neuro-plasticity is the technical term used to describe a very interesting property of the nervous system. It means that the nervous system can and does change or remodel. This remodeling and changing to the nervous system occurs when signals traveling within the nervous system become interrupted or distorted.

Peripheral neuropathy is a disease in which changes and interruptions in nerve signals is widespread. Changes and remodeling of the nervous system and brain occur frequently in patients suffering from neuropathy, as we described in earlier sections of this paper.

Refer back to figure 5, the human homunculus of the brain. If you look closely, the brain has specific areas that correspond to and represent different parts of the body. There is a specific area for the face, trunk, arms, legs, and every other part of the body. In our drawing, the area representing the face is probably easiest to identify. The area that represents the legs is marked with a "V" in this illustration (figure 5).

Studies of the brain known as functional magnetic resonance imaging, or fMRI for short, have documented which parts of the brain are associated with each body region. fMRI studies are very useful in demonstrating neuro-plasticity. [122-124]

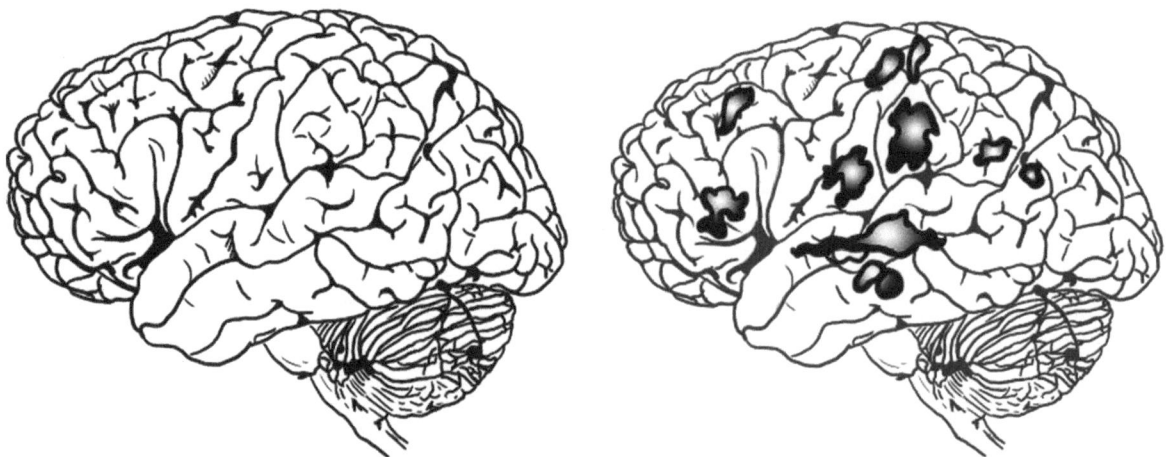

Figure 18, Illustrates how specific areas of the brain "light-up" as a result of specific stimulation on various body parts. The hands, feet and other body parts are said to be represented in highly specific maps of the brain. This technique called functional magnetic resonance imaging or fMRI can demonstrate the effects of a nerve injury in the periphery on the brain. It can also be used to demonstrate the response of the nervous system to specific treatments. The image above left is a fMRI of the brain before acupuncture. The image on the left shows increased activity is specific brain regions after acupuncture. It thus appears that acupuncture may be able to restore the normal patterns of brain activity. [122-124]

For example, these studies show that when a patient has carpal tunnel syndrome in the wrist and the nerve signal from the wrist to the brain is altered, the corresponding part of the brain is affected. What these studies show is whenever the nerve signal fails to be transmitted from a body part to the corresponding region of the brain, the specific brain region becomes blurred. It starts to blend with surrounding regions and loses its association with its peripheral body part. The representation of the body part on the surface of the brain becomes fuzzy and blended. This process is called neuroplasticity.

When this blurring occurs, it is believed to represent both chemical and structural changes within the brain.

Recent studies show that acupuncture given for as little as five weeks can help to reverse this neuroplasticity in the brain and restore the normal patterns seen in the brains of healthy individuals [125-127].

What is even more remarkable is the fact that these brain changes in response to acupuncture only occur when needles are inserted into true acupuncture points. Random insertion of needles does not produce the same response as seen in stimulation of actual acupuncture points [126].

Furthermore, the changes in brain neuroplasticity caused by acupuncture do not occur in healthy individuals that were acting as control subjects in these experiments [125].

Most of the published studies of brain neuroplasticity have been conducted on patients who have a single nerve affected, like carpal tunnel syndrome patients.

Carpal tunnel syndrome, even though it only interferes with the signals traveling from the median nerve in the wrist to the brain, still causes altered metabolism and neuroplasticity in the brain. [126] Imagine for a minute the massive changes that must occur in the neuropathy patient's brain.

In peripheral neuropathy, most, if not all, of the nerves in the body are affected. The interference with the nerve signals from various body parts to the brain is widespread.

The changes and neuroplasticity in the brain of a neuropathy patient are substantial. This is why treatment of the patient suffering from neuropathy must be comprehensive.

Acupuncture treatment in particular neuroanatomical acupuncture, based on the most modern techniques of neurology, is a very promising and indispensable component of the comprehensive management of patients suffering from peripheral neuropathy.[119-121, 128-134]

Section II: The neurobiology of alternative therapy for the treatment of peripheral neuropathy

Introduction

As you may have already realized from previous sections of this manuscript, the nervous system is made up of extremely complex biological tissues. So trying to help patients suffering from neuropathy is challenging. The task is made even more difficult because despite our understanding of how nervous tissue functions, we still do not have a complete understanding of the biology of the nervous system. We are in effect, trying to put together a very complex jig-saw puzzle, without having all the pieces. In this section we will take a look at some of the known biology of the nervous system and how it becomes affected in patients suffering from neuropathy. Then suggest some natural substances with potential to correct these neurological abnormalities. Keep in mind our understanding of these biological systems is rudimentary and how these systems interact to contribute to symptoms of neuropathy is not well understood. None the less, we do have some opportunities to try to influence the pathology of neuropathy and produce meaningful benefits for our patients.

Much of our understanding of the neurobiology of neuropathy has been derived from studies of diabetes. How much of this knowledge can be extrapolated to other types of neuropathy is subject to debate.

There are several broad based therapeutic approaches when considering the treatment of neuropathy. The first is to treat the neuropathy symptomatically. Another approach discussed below is to try to bypass the defective signaling in the nervous system. This approach attempts to carry on the lost function of neurons "downstream" of the damage or impairment. In essence by-passing the loss of function at the site of the damage to the neuron. The third approach is to attempt to rebuild nerve damage in patients suffering from neuropathy. A number of natural substances show promise in this regard. Some of these natural substances have already been subject to clinical trials in neuropathy patients and have demonstrated real therapeutic benefit. The separate approaches are not mutually exclusive. The approach used in our office attempts to quell the symptoms of neuropathy while also attempting to regenerate nerves and restore lost function either directly or through by-pass of lost function.

There are a number of basic science studies that suggest that natural substances can modulate: 1. Sodium and calcium channels, 2. Protein Kinase C, 3. Aldose Reductase, 4. Insulin signaling, 5. Neuro-trophic Growth Factors or 6. GABA-nergic neurotransmitter activity and may be potential treatments for patients suffering from neuropathy. Some of this research is in the basic science stages; other data is derived from clinical trials.

Several promising avenues of treatment that are in various stages of therapeutic clinical application include sodium ion channels blockers [14, 135-137], calcium channel blockers [4, 138-140], Protein Kinase C inhibitors (PKC)[141-152], Thiamine [153-157], Aldose Reductase Inhibitors [154, 158-161], direct insulin signaling modulators [162-166] and neurotrophic substances.[167-178]

The role of each of these biological systems in the development of and as a potential treatment for neuropathy has been derived from basic science, animal, human and some preliminary human clinical drug trials.

We have previously discussed the concepts of neuroplasticity and the role of ion channels in the development of neuropathy. Recent studies have demonstrated that a highly specific type of neuroplasticity occurs in the axon of nerves in neuropathy patients.[179] [137] The axons actually remodel and redistributes ion channels in the axon cell membrane. This redistribution of ion channels increases the density of channels, specifically of sodium channels. [14] The net effect of this change in the density of ion channels is facilitation of the nerve axon. [14] [137]This means pain signals can more easily be transmitted from the periphery, like the feet and legs, up to the brain.

Very much like sodium channels, calcium channels also affect the ability of an axon to transmit electrical signals. The state of calcium channels also plays a role in the facilitation or suppression of electrical activity in the nerve axon. Sodium, calcium and other ion channels all influence the state of a nerve axon. Changes in these ion channels, render the neuron more or less capable of transmitting pain and other signals along the course of the nerve.

Direct and indirect evidence supports the role of calcium channel blockade in the suppression of the painful electrical signals. There are a number of different types of calcium channels and research suggests that N & T type calcium channels play a role in neuropathy and chronic pain expression.

In fact the two drugs commonly prescribed for the treatment of neuropathy, pregabalin and gabapentin, appear to work in part through their ability to block calcium channels in the nervous system.[4, 5, 139]

At least in the animal model of diabetic neuropathy, activation of calcium channels is associated with the development of hypersensitivity and allodynia. [17, 180] The T-type calcium channel appears to be an amplifier of painful sensory stimuli. [16]

In mice genetically bred to lack N-type calcium channels, decreased pain perception was demonstrated and the development of neuropathic pain was inhibited suggesting the participation of this sub type of calcium channel in neuropathic pain syndromes.[18] Ziconotide, a selective N-type calcium channel blocker was reported to ameliorate intractable neuropathic pain that failed to respond to either direct spinal cord stimulation or morphine infusion. [15]

The antiepileptic drugs are showing promise in relieving neuropathic symptoms. First-generation antiepileptic drugs have been shown to be effective in neuropathic pain. The evidence supporting the use of a new generation of antiepileptic drugs in painful diabetic neuropathy has been reviewed.[5] Lyrica® or pregabalin and Neurontin® or gabapentin are two drugs currently FDA approved to treat neuropathies. At least part of the mechanism of action of these drugs appears to be modulation of calcium influx into the neuron through the regulation of the calcium channels.

Pregabalin (Lyrica) is an alpha(2)-delta ligand that binds to and modulates voltage-gated calcium channels, exerting its intended effect to reduce neuropathic pain. [4]

Gabapentin (Neurontin), at clinically relevant concentrations, results in significant reduction of intracellular calcium I(Ca) in both control and neuropathic neurons.[139]   Neurontin's depression of intracellular calcium may be partly related to the binding of the drug to the alpha(2)delta modulatory subunit of the voltage activated calcium channels. [139]   Neurontin's analgesic effect then is likely due to diminished release of neurotransmitters by sensory neurons which is known to be a calcium dependent process.[139]

Taken together it appears that calcium channels and in particular N and T type calcium channels play a role in the development of neuropathic pain.   Thus calcium channel blocking agents appear to have a role in the treatment of nerve related pains like neuropathy. Later in this chapter we will discuss natural substances that have demonstrated calcium channel blocking properties.

Many of the drugs used to "numb" tissue like Novocain, work by blocking sodium channels.   Blocking the sodium ion channels in a nerve greatly reduces or completely blocks its ability to carry electrical impulses. [14, 140, 180]   This is how a dentist numbs your mouth to prevent you from feeling pain during dental procedures.   Agents that block sodium channels are being considered as potential treatments for neuropathic pain.[14, 137]   There are several natural substances that appear to be able to block sodium channels. This makes them biologically plausible topical or internal agents in the treatment of neuropathy symptoms.[181]

PKC

Protein kinases are enzymes which catalyze the transfer of phosphate from adenosine-5'-triphosphate (ATP) to certain amino acid residues in specific proteins. Generally, the phosphorylation of a protein changes its functionality, from inactive to active in some cases, and from active to inactive in others. Thus the mechanism through which signaling pathways are switched on or off is through phosphorylation. Phosphorylation is a means of up-regulating or down-regulating processes controlled by specific proteins.

There are a number of different forms of these enzymes that demonstrate diverse biological affects in tissue. At least one of these forms, protein kinase C, may play a role in the development of diabetic neuropathy [141, 142, 145, 146, 150-152]. Protein kinase C or PKC for short, appears to be overactive in diabetic patients. [145] Drugs that block PKC appear to positively influence many of the complications of diabetes including neuropathy [143, 147, 148, 182-185]

Protein Kinase C and calcium and sodium channels in neurons interact and this interaction appears to play a role in neuropathic changes in the nerve axon.[150] [141, 146] PKC activation is known to contribute to C fiber (pain carrying nerve fiber) activity. [183]

PKC & Neuropathy:

It is known that Protein Kinase C activity contributes to C-fiber afferent excitability.   The effects of agents that inhibit protein kinases on behavioral mechanical nociceptive thresholds and on the response of C-fiber afferents to sustained mechanical stimulation have been studied [183].

Agents that inhibit PKC appear to increase the mechanical nociceptive threshold (pain thresholds) of diabetic rats in a dose-dependent manner, but do not alter nociceptive threshold in control, non-neuropathic, rats.[183]

Investigators have found that both the mechanical behavioral hyperalgesia and the C-fiber hyperexcitability to mechanical stimuli seen in diabetic rats are reduced by agents that inhibit Protein Kinase C. This evidence supports the hypothesis that C-fiber hyperexcitability is at least in part mediated by PKC activity and contributes to hyperalgesia in this particular model of diabetic neuropathy. [183]

There are a number of other studies the results of which support the association of alteration in neuronal PKC activity and the development of the signs and symptoms of neuropathy.

Some additional research results that corroborate this association include the finding that PKC is involved with the development of neuropathic pain.

In one particular study, it was suggested that PKC interacts with a specific type of sodium channel in the neuron. The researchers observed no pain related behavior in genetically modified mice that lack the specific sodium channel, known as TRPV1.

Thus acute pain related behavior after injection of phorbol 12-myristate 13-acetate, an activator of protein kinase C (PKC), into the paw of experimental animals, was not seen in mice bred to lack the TRPV1 sodium channel receptor. These observations suggest that PKC activation elicits pain through a sodium channel receptor dependent mechanism via either sensitization or activation of the TRPV1 receptor. [186]

Furthermore, elevated protein kinase C (PKC) activity is thought to play a substantial role in the etiology of diabetic microvascular complications. The PKC-beta isoform has been identified as particularly important contributor to diabetic complications. Since neuropathy has a vascular component one study sought to assess whether a PKC inhibitor could correct nerve conduction velocity (NCV) and perfusion abnormalities seen in the nerves of diabetic rats. [185]

The results seem to indicate when combining a PKC inhibitor with doses of the antioxidants vitamin E, alpha-lipoic acid, or Gama Linoleic Acid (GLA) in diabetic rats, motor and sensory nerve conduction velocities and sciatic nerve perfusion were restored to within the normal non-diabetic range. The combined effect was equivalent to that of high doses of PKC inhibitor administered as a sole treatment. These findings appear to demonstrate synergism between PKC-beta inhibition, oxidative stress and essential fatty acid mechanisms.[185]

More evidence that PKC contributes to the complications of Diabetes is reported. Researcher have also found that the diabetic state leads to protein kinase C (PKC)-beta over-activation and microvascular dysfunction, possibly resulting in disordered skin microvascular blood flow (SkBF) and other changes observed in diabetic peripheral neuropathy (DPN) patients[143]

Endothelium-dependent and C fiber-mediated SkBF, sensory symptoms, neurological deficits, nerve fiber morphometry, quantitative sensory and autonomic function testing, nerve conduction studies and quality of life were examined in diabetic patients. [143]

PKC hyperactivity can be achieved through the administration of the drug, 1alpha,25(OH)2D3 which induces activation of PKC alpha and in turn enhances extra-cellular calcium entry into the cell.[141] Suggesting that PKC over activation produces cell damage at least in part through calcium related mechanisms.

Substances that are known to inhibit PKC have been shown to be beneficial in patients suffering from diabetic complications.

In a cohort of neuropathy patients, ruboxistaurin (a potent PKC inhibitor) was shown to enhanced skin blood flow at the distal calf, reduced sensory symptoms, improve measures of Norfolk neuropathy quality of life scale and was well tolerated.[143]

Significant improvements from baseline within the ruboxistaurin (PKC inhibitor) group were also observed for the Neuropathy Total Symptom Score-6, the Norfolk QOL-DN symptom subscore and total symptom score reduction (average reduction 41.2%)[143]

Chronic hyperglycemia is a major initiator of diabetic micro- and cardiovascular complications, such as retinopathy, neuropathy and nephropathy. Several hyperglycemia-induced mechanisms may induce vascular dysfunctions, which include increased polyol pathway flux, altered cellular redox state, increased formation of diacylglycerol (DAG) with the subsequent activation of protein kinase C (PKC) isoforms and accelerated non-enzymatic formation of advanced glycinated end products,(AGEs) [145]

Hyperglycemia alters pain sensitivity. In humans, diabetic neuropathy can be associated with burning and tactile hypersensitivity. Behavioral reactions of hyperalgesia in animal models of diabetes have been described. However, the etiology of these disturbances is still unknown. Metabolic factors such as hyperglycemia or neurotransmitter alteration may be involved. Activation of protein kinase C (PKC) has been implicated in changes in pain perception commonly found in diabetic neuropathy patients.[148]

Electrophysiological experiments have shown that activation of PKC leads to long-lasting enhancement of excitatory amino acid-mediated currents in dorsal horn neurons and trigeminal neurons. Thus, activation of PKC may underlie the neuronal sensitization that produces hyperalgesia in diabetic neuropathy.[148]

PKC Inhibitors:

A number of natural Protein Kinase C inhibitors have been identified. Refer to the matrix in *table 1* to see natural substances and their biological activities relative to ions channels, PKC, and neuropathy.

Ginko

Several studies suggest Ginko Biloba may influence PKC activity. Data from one particular study suggest that the protective and rescuing abilities of EGb 761, a ginko derivative, are not only attributable to the antioxidant properties of its flavonoid constituents, but also via their ability to inhibit NO-stimulated PKC activity.[187]

The ginko derivative, EGb 761 and red wine-derived polyphenols were found to protect neurons due in part to their antioxidant activities and also because of their ability to block SNP-stimulated activity of protein kinase C (PKC). Taken together, these results support the hypothesis that dietary intake of natural substances that may inhibit the activation of PKC, may be beneficial in normal aging of the brain. [188]

Vitamin E

From in vitro studies, a large amount of data has shown that alpha-tocopherol (the major component of vitamin E) regulates key events in the cellular pathogenesis of atherosclerosis. Researchers have identified the inhibition of protein kinase C (PKC) activity by alpha-tocopherol as the basis of the vascular smooth muscle cell growth inhibition associated with vitamin E compounds [189]

Researchers have concluded that the effect of alpha-tocopherol against atherosclerosis is not due only to the prevention of LDL oxidation, but also due to the down regulation of the scavenger receptor CD36 and to the inhibition of PKC activity.[189]

In addition, high doses of vitamin E were shown to decrease the level of DAG and subsequently PKC hyperactivity induced by diabetes or hyperglycemia. Thus animal and clinical studies have shown that high doses of vitamin E treatment can apparently reverse some of the changes in the retinal and renal vessels seen in association with hyperglycemia.[184]

In mammary tissue researchers demonstrated that the inhibitory effects of specific vitamin E derivatives on normal mammary epithelial cell growth and division, occurred downstream from the EGF-receptor and appears to be mediated, at least in part, by a reduction in PKC(alpha) activation.[191] These and other findings further suggest that vitamin E (alpha-tocopherols) show promise for reversal of PKC activation.[190]

Aldose Reductase Inhibitors

Aldose reductase is an enzyme that is involved with the metabolism of sugars and sugar alcohols. [158, 159] It has been found to be important in the development of diabetic complications including neuropathy. [158, 159] Substances that block this enzyme, known as aldose reductase inhibitors, show promise in the prevention and treatment of diabetic complications such as diabetic neuropathy. Their clinical usefulness has been limited, however, because most synthetic aldose reductase inhibiting agents produce unwanted and unacceptable side effects. However, simple vitamin C and some of its derivatives appear to be safe and natural aldose reductase inhibiting agents. [158, 192-194] Likewise the flavinoid quercetin is a potent inhibitor of the aldose reducatase enzyme and may play a role in the management of diabetic neuropathy. [195-200]

Thiamine

The B vitamin Thiamine (B1) participates in many metabolic processes and has important implications for the treatment of neuropathy. A deficiency of B1 itself will cause neuropathy, but it also appears that mega doses of B1 can be used to alter a number of metabolic pathways that may prevent or reverse certain types of nerve dysfunction. [153, 157, 201-205]

Lipid soluble B1, Benfotiamine or Allithiamine [206] are thought to be more biologically active forms of thiamine. Allithiamine has been shown to be absorbed as rapidly as parentally (intravenous) administered water soluble B1. [207] In the case of benfotiamine, preliminary studies suggest it can improve symptoms of neuropathy. [154, 208-214]

The therapeutic benefits of B1 supplementation in neuropathy may be do to its effects on the accumulation of triosephosphates arising from the high glucose concentrations in diabetes. These triosephosphates are a likely or potential trigger for many of the biochemical abnormalities that lead to the development of diabetic complications. Some of these metabolic abnormalities may be prevented or reversed by removal of excess triosephosphates via a metabolic cascade known as the pentosephosphate pathway. This pathway is impaired in experimental and clinical diabetes by even mild thiamine deficiency. The activity of transketolase, the pacemaking enzyme in this complex chemical cascade is thiamine-dependent. Supplementation with thiamine may therefore activate the transkeltolase enzyme within the pentosephosphate pathway and thus reduce or reverse the accumulation of triosephospahte and its associated harmful effects in diabetic patients. [157]

Oral supplementation with allithiamine normalizes transketolase activity in red blood cells.[207] Likewise benfotiamine has a demonstrated ability to affect this pathway in a clinically meaningful way. [208]

Other studies suggest that thiamine in its lipid soluble form may positively influence the metabolic pathways that become deranged as a result of excess glucose accumulation.

Elevated blood glucose leads to glycination of proteins. This process is believed to cause degradation of proteins and tissues. Advanced Glycination End-products (AGEs) are considered to be major players in tissue degeneration and thus complications of diabetes. [153]

The formation of AGEs is suppressed by tight and aggressive blood sugar control. AGEs may also be suppressed by thiamine, pyridoxamine supplementation and several other pharmacological agents. Increasing expression of enzymes that are part of the enzymatic defense against glycation provides a novel and potentially effective future therapeutic strategy to suppress protein glycination and its harmful affects.[153]

High glucose also induces pathological alterations in small and large blood vessels, possibly through increased formation of advanced glycination end products (AGE), activation of aldose reductase and Protein Kinase C, and increased flux through the hexosamine pathway.[154] These biochemical abnormalities are to various degrees related to the development of diabetic complications like neuropathy.

Thiamine and benfotiamine also correct increased apoptosis (cell death)due to high glucose in cultured vascular cells. Further elucidations of the mechanisms through which they work could help establish the basis for clinical use of these vitamins in the prevention and or treatment of diabetic microangiopathy.[154]

Dysfunction of beta-cells and impaired glucose tolerance in thiamine deficiency and the observed link between impaired glucose tolerance with dietary thiamine, suggests that thiamine therapy may have a future role not only in the prevention of diabetic complications, but also in prevention of type 2 diabetes itself. [157]

Neurotrophic substances

The discovery of substances that stimulate nerve growth and regeneration, also known as neurotrophic factors, has lead to great excitement in the field of neurology. [175, 177, 215]

Prior to the discovery of these chemicals, damaged nerves were thought to lack the capability of regeneration and repair. The discovery of substances that promoted nerve sprouting and re-growth changed our understanding of the nervous system, nerve injury and nerve repair. Excerpts from various studies of neurotrophic substances are discussed below.

Preclinical and ongoing clinical trials of nerve growth factor (NGF) suggest that it will be useful for the treatment of diabetic, toxic and compressive sensory neuropathies. At appropriate doses NGF has no significant negative side effects in humans.

Other preclinical studies suggest that insulin-like growth factor 1 (IGF-1) will be useful for the treatment of mixed motor and sensory neuropathies. For example, IGF-1 treatment can prevent the experimental motor and sensory neuropathies caused by the anti-tumor drugs, vincristine and cisplatin. [167, 168]

It was observed that both IGF-I and EGb761 (a Ginko derivative) are unique in that they are able not only to protect, but even to rescue neurons against A beta toxicity.[216]

Another study determined that insulin and insulin-like growth factors (IGFs) have neurotrophic actions on sensory, sympathetic and motor neurons. These are the main types of neurons afflicted in diabetic peripheral neuropathy. Moreover, it is interesting to note that IGF activity is reduced in both clinical and experimental diabetes. [171]

The authors of this study report that (i) IGF activity is reduced in diabetic neural tissues; (ii) conduction velocity is impaired in the diabetic spinal cord; (iii) replacement therapy with IGF can prevent neuropathy in diabetic nerves; and (iv) IGFs can prevent diabetic neuropathy, despite ongoing hyperglycemia. [171]

A review of the literature concerning neurotrophic substances suggests that motor neuronal disorders, such as the loss of spinal cord motor neurons in amyotrophic lateral sclerosis or the degeneration of spinal cord motor neuron axons in certain peripheral neuropathies, present a unique opportunity for therapeutic intervention with neurotrophic proteins.

Normally, such proteins do not cross the blood-brain barrier, but spinal cord motor neuron axons and nerve terminals lie outside the barrier and thus may be targeted by systemic administration of protein growth factors. Insulin-like growth factor-I (IGF-I) receptors are present in the spinal cord [173]

Consistent with other reports that IGF-I enhances motor neuronal sprouting in vivo, subcutaneous administration of IGF-I increases muscle endplate size in rats. Subcutaneous injections of IGF-I also accelerates functional recovery following sciatic nerve crush in mice, as well as attenuates the peripheral motor neuropathy induced by chronic administration of the cancer chemotherapeutic agent vincristine in mice.[173]

Insulin-like growth factor-1 (IGF-1) and its receptors share considerable homology with insulin and insulin receptors, and their respective signaling pathways interact at the post receptor level. Experimentally, IGF-1 has a protective effect on neuropathy, and could find an application in the healing of neuropathic ulcers. The potential benefits of IGF-1 therapy in diabetes mellitus have yet to be realized.[215]

So these nerve growth factors show definite promise in the treatment of neuropathy. That being said, there are a number of natural substances that may be employed to either directly stimulate nerve growth and repair or that may stimulate the release of neurotrophic substances. They may have potential in the overall comprehensive management of patients suffering from peripheral neuropathy. Some of these substances in addition to exhibiting neurotrophic activity also have other properties that may prove beneficial to the neuropathy patient. They will be discussed in this section and elsewhere in this report.

The most interesting of the natural substances has a history of use in Traditional Chinese Medicine that dates back thousands of years. It was considered a nerve tonic and rejuvenator. Recent scientific studies have demonstrated that it may prove useful as a stimulant for nerve regeneration and repair.

Elk Velvet

The antlers from deer drop seasonally and then re-grow. What makes this phenomenon so remarkable is the fact that the growing antlers have within them peripheral nerves. Nowhere else in nature do mammalian nerves grow at such an extraordinary rate. Recent research confirms that deer antler velvet contains large amounts of nerve growth factors and neurotrophic substances [170, 176, 217-219] In Chinese Medicine antler velvet has a long history of use as a nerve tonic. The potential of using antler velvet as a natural source of nerve growth factors and IGF is intriguing. Antler velvet then would seem to have potential in the management of neuropathies.

Gotu Kola

A single study reports that the herb Gotu Kola may accelerate nerve sprouting and repair. [220] The mechanism of the observed accelerated growth appears to be associated with the increased production of nerve growth factors caused by consumption of Gotu Kola. [220]

Vanadium.

Later in this paper we will discuss the possible application of vanadium compounds in the management of neuropathy. Vanadium mimics the action of insulin and may also enhance the effects of insulin like growth factor either directly or indirectly. [221-223]

Direct insulin signaling:

The basic model of nerve damage in diabetic neuropathy has centered around excessive blood sugar and abnormalities in sugar metabolism as the cause of neuropathy. However a number of studies suggest that lack of insulin itself or lack of the signaling and metabolic activation attributable to insulin, may also be important to the development and continuation of neuropathy.[162-165] Researchers have found that C-peptide, a cleavage product of endogenous insulin production, exerts a dose dependent protection against type 1 diabetic neuropathy in rats. [166]

This is an interesting avenue to consider. It lends itself to several natural substances that may be used to stimulate the insulin receptor directly or by-pass the insulin receptor and activate or enhance downstream metabolic pathways associated with insulin.[224-226]

Vanadyl Sulfate and its derivates.

A little over one hundred years ago, vanadium-containing compounds were assessed clinically for use in treatment of human diabetic patients.[227] More recently, it has been found that some vanadium derivatives display potential as insulin mimicking substances. Oxovanadium(IV) complexes have potent insulin-like activity and have been shown to be compounds capable of treating both type I and type II diabetes in animals. [224]   Among these related compounds, VO(3mpa)(2) was found to be the most potent activator in inducing not only the phosphotyrosine levels of both IRbeta and IRS, but also the activation of downstream kinases in the insulin receptor, such as Akt and GSK3beta, which in turn activated the insulin-dependent GLUT4 enzyme in the plasma membrane.[224] The present data from this study indicate that both activation of insulin signaling pathways, which follows the GLUT4 activation and enhancement of glucose utilization by oxovanadium(IV) complexes cause the normalization of elevated blood sugar levels seen in diabetic animals. [224]

The different derivatives of vanadyl sulfate exerted variable affects on the diabetic condition. Some compounds demonstrated higher insulin-like activity compared with VOSO4, other vanadyl compounds had activity significantly lower than that of VOSO4. These results indicate that specific derivatives of VOSO4 complexes might be a promising approach to obtain superior insulin-like activity.[225]

At the cellular level, vanadium activates several key elements of the insulin signal transduction pathway, such as the tyrosine phosphorylation of insulin receptor substrate-1, and extra-cellular signal-regulated kinase 1 and 2, phosphatidylinositol 3-kinase and protein kinase B. [226] These and similar results raise the possibility that vanadium derivatives may serve not only as an adjunctive therapy for lowering blood glucose levels, but may also stimulate many of the downstream metabolic pathways that in theory, may prevent, delay or reverse diabetic neuropathy. This makes vanadium mineral derivatives exciting potential natural substances in the treatment of diabetes and its neurological complications.

Alpha Lipoic Acid:

Alpha Lipoic Acid (ALA) is a unique and potent antioxidant substance. It is both lipid and water soluble. Studies have suggested that alpha-lipoic acid, a potent antioxidant, improves symptoms of diabetic neuropathy. [228] Clinical studies also demonstrate both the efficacy and safety of alpha lipoic acid therapy in patients with diabetic neuropathy.[229] While the exact mechanism of the benefits of ALA are unknown, alpha-lipoic acid has been shown to acutely stimulate the insulin-signaling cascade, thereby increasing glucose uptake in muscle and fat cells.[230]

ALA effectively improves the sensory symptoms of diabetic polyneuropathy and is safe for most diabetic patients. [231] Alpha Lipoic Acid has been proven to be an effective, safe, and cost-effective treatment option for the majority of patients suffering with diabetic complications particularly neuropathy. [232]

One study found that an alpha-lipoic acid preparation, in dosage 600 mg daily for 3 months resulted in the significant reduction of clinical (sensory and motor) and neurological changes of lower extremity peripheral nerves. [233] Another study demonstrated oral treatment for 4-7 months tended to reduce neuropathic deficits and improve cardiac autonomic neuropathy. [234]

Alpha lipoic acid was compared with gabapentin, the mainstay of medical management of neuropathy. The study demonstrated that switching from long-term treatment with alpha-lipoic acid to central analgesic drugs such as gabapentin in painful diabetic neuropathy was associated with considerably higher rates of side effects, frequency of outpatient visits, and daily costs of treatment. These results indicate that alpha lipoic acid may be a cost effective, safe and clinically useful alternative to gabapentin for patients suffering from neuropathy.

Based on these results, alpha lipoic acid (ALA) should be considered as a treatment option for patients with peripheral diabetic neuropathy. Clinical and post-marketing surveillance studies have revealed a highly favorable safety profile of this antioxidant [234]. When discussing supplementation with patients, it is important to discuss potential side effects, vitamin, mineral, and drug interactions and current evidence available regarding efficacy. [235]

Ginseng:

Ginseng extracts produce analgesia among other various biologically beneficial effects. [236] The pain suppressing effects induced by ginsenosides are dose dependent [237]

Research findings strongly suggest that the ginseng saponins, especially ginsenoside Rg2, block the nicotinic ACh receptors or the receptor-operated Na+ channels (but not voltage-sensitive Na+ and Ca++ channels), inhibit Na+ influx through the channels and consequently reduce both Ca++ influx and catecholamine secretion in bovine adrenal chromaffin cells. [238]

Ginseng & Calcium Channels

It is known that there are at least five different Ca(2+) channel subtypes in neuronal cells[239] Previous studies have reported that ginseng derivatives inhibit high threshold voltage-dependent Ca(2+) channels in neuronal tissues[239]

This study reported that ginsenosides inhibited high threshold voltage-dependent Ca(2+) currents in a dose-dependent manner.[239] The data suggest that ginsenosides are coupled to three types of calcium channels, including the (N-type) channel, (P-type) channel and (presumptive Q-type) channel.[239] Another study reported that ginseng and it's constituent ginsenosides exerted inhibitory effects on Ca2+ currents in rat adrenal chromaffin cells.

These results suggest that ginseng saponins regulate catecholamine secretion from adrenal chromaffin cells and this regulation could be the cellular basis of antistress effects induced by ginseng. [240]

The ginseng constituents ginsenosides Rf and Rg(3) have been identified as active components in ginseng-mediated neuroprotection. The results of this study suggest that inhibition of L-type Ca(2+) channels by ginseng could be one of the mechanisms for ginseng-mediated neuroprotection in cultured rat cortical neurons. [241]

Several studies found ginseng exhibited properties similar to opioids. One study reported panax ginseng root extract acts on sensory neurons through a similar pathway as mu-type opioids. Both ginseng and opioids inhibit Ca2+ channels through pertussis toxin-sensitive GTP-binding proteins. However, the receptor for ginseng root extract is not an alpha 2-adrenergic, GABA~B, muscarinic, or opioid receptor [242] Another study found that at saturating concentrations, ginseng derivatives rapidly and reversibly inhibits N-type, and other high-threshold, Ca2+ channels in rat sensory neurons to the same degree as a maximal dose of opioids [243]

Ginseng and Sodium Channels

As we discussed repeatedly in this paper, voltage-gated Na(+) channels in primary sensory neurons play important roles in pain perception. [236] There are many components of ginseng that have diverse biological effects. One of the constituents of ginseng is EHD. [236] EHD is interesting in that it was shown to inhibit Na(+) currents in a concentration-dependent manner. [236] Other attributes of this ginseng derived compound include, accelerated inactivation of both Na(+) currents and a hyperpolarizing shift of the steady-state inactivation curve. In addition EHD suppressed the maximal Na(+) current at negative holding potentials at which the channels are relieved from inactivation.[236] Taken together these findings suggest that specific components of ginseng may have potential for suppressing abnormal sensory nerve activity in neuropathy patients through modulation of sodium ion channel mechanisms.

Neuronal damage during ischemic episodes has been associated with abnormal Na(+) fluxes. Drugs that block voltage-dependent Na(+) channels provide protection to nerve cells during cerebral ischemia [244] This study investigated the effects of American Ginseng extracts and compared them with the effects of the known sodium channel blocker, lidocaine.

The results suggest that Na(+) channel block by American Ginseng extract was primarily due to interaction with the inactive state of the channel. The investigators concluded that inhibition of the Na(+) channel activity by American Ginseng extract may contribute to its neuro-protective effect during ischemia. [244]

Ginseng and GABA Receptors

In this study it was demonstrated that ginseng regulated GABA(A) receptor expression in Xenopus oocytes and suggests that this regulation might be one of the pharmacological actions of Panax ginseng [245]

Acetyl Carnitine

Carnitine is an amino acid. Amino acids are the building blocks of proteins. A special form of the amino acid is known as acetyl carnitine. Acetyl carnitine seems to be more easily absorbed and available to the nervous system than the non acetylated form of carnitine. Acetylated carnitine has been studied extensively for its ability to protect and repair damage within the nervous system. It appears to be a very promising natural therapy for various types of neuropathies. In this section we will review some of the studies of acetyl carnitine.

In animal models of painful peripheral neuropathy due to nerve trauma or diabetes there is obvious axonal degeneration accompanied by an abnormal incidence of spontaneous discharge in A-fiber and C-fiber pain related neurons. [250]

In addition to it's known ability to stimulate nerve re-growth and repair, Acetyl L Carnitine, or ALC is known to produce a strong pain relieving effect when given after neuropathic pain has already been established. ALC appears to also improve the function of peripheral nerves by increasing nerve conduction velocity, reducing sensory neuronal loss, promoting nerve regeneration and stimulating neuroplasticity throughout the pain pathways. [246]

Acetyl-L-carnitine (ALC), the acetyl ester of L-carnitine, plays an essential role in intermediary metabolism. Some of the properties exhibited by ALC include neuroprotective and neurotrophic actions, antioxidant activity, positive actions on mitochondrial metabolism, and stabilization of intracellular membranes. ALC has demonstrated efficacy and high tolerability in the treatment of neuropathies of various etiologies, including chemotherapy-induced peripheral neuropathy (CIPN). [247] Additionally, ALC has gained clinical interest for its analgesic effect in different forms of neuropathies associated with chronic pain, such as diabetic and HIV-related peripiperal neuropathies. [246]

ALC administration promotes the recovery of nerve conduction velocity, restored the mechanical nociceptive threshold, and induced analgesia by up-regulating the expression of type-2 metabotropic glutamate receptors in dorsal root ganglia. [247]

In animal studies, treatment with acetyl-L-carnitine, but not with the common form of this amino acid (L-carnitine) or the approved neuropathy medication Gabapentin, was shown to prevent apoptosis (death of the nerve cell) [248]

Demonstrating that ALC is able to prevent regulated cell death and loss of nerve cells in damaged sciatic nerves. [248]

ALC is a novel therapy currently being widely studied as a treatment for peripheral neuropathies.[249]

Carnitine & Neuropathy

The antinociceptive (pain relieving) effect of ALC has been confirmed in several experimental models of neuropathic pain, including streptozotocin induced, chemotherapy-induced and the sciatic nerve chronic constriction injury models of neuropathy. In these models, prophylactic administration of ALC has proven to be effective in preventing the development of neuropathic pain. [246]

Chemotherapy Induced

In the nervous system, the major brunt of the toxicity associated with chemotherapy is directed against the peripheral nerve, resulting in chemotherapy-induced peripheral neuropathy (CIPN). [251]

The development of neuropathy in chemotherapy patients may be related to the initiation of spontaneous discharges in the fibers associated with pain signaling. The affects of chemotherapy on nerve conduction activity have been reported.

As compared with vehicle-injected placebo control animals, researchers found a significant increase in the number of spontaneously discharging A-fibers and C-fibers in animals given chemotherapy agents. As has been discussed these fibers are associated with pain transmission. Moreover, they have shown that prophylactic treatment with acetyl-l-carnitine (ALC), which is known to block the development of the paclitaxel-chemotherapy evoked pain, causes a significant decrease (~ 50%) in the number of A-fibers and C-fibers with spontaneous discharge.

These results suggest that abnormal spontaneous afferent nerve fiber discharge is likely to be a factor in the genesis of chemotherapy-evoked painful peripheral neuropathy, and that the therapeutic effects of ALC may be due to the suppression of this abnormal spontaneous discharge activity. [250]

Other preliminary studies of chemotherapy induced neuropathy have shown promise for several natural agents including glutamine, glutathione, vitamin E, acetyl-L-carnitine, calcium, and magnesium infusions.

However, final recommendations for the use of these nutritional treatments in chemotherapy induced neuropathy await additional prospective confirmatory studies. [251]

A review of the available literature focusing on the clinical features, mechanisms, and possible therapy for the neurotoxicity of chemotherapy has been published. In particular, oxaliplatin, thalidomide, methotrexate, ifosfamide, cytarabine, amifostine, acetyl-L-carnitine, methylene blue, cytokines, and neurotrophins were reviewed and discussed.[252]

Several of the more remarkable reported findings include: Sensory neuropathy grade improved in 15 of 25 (60%), and motor neuropathy improvement in 11 of 14 patients (79%). The total neuropathy score including neurophysiological measures improved in 23 (92%) of the patients receiving ALC compared with controls.

These symptomatic improvements persisted in 12 of 13 patients for a median of 13 months after ALC supplementation. The authors conclude that in view of ALC's effect in improving established paclitaxel and cisplatin neuropathy, they recommend ALC treatment in preventing progression or even in the reversal of symptoms which occur as a result of neurotoxic chemotherapy.[253]

In a separate experiment, daily administration of ALC (100 mg/kg; p.o.; for 10 days) to rats with established paclitaxel-induced pain produced an analgesic effect. This effect dissipated shortly after ALC treatment was withdrawn. The authors conclude that ALC may be useful in the prevention and treatment of chemotherapy-induced painful peripheral neuropathy. [254]

Furthermore, after the administration of neurotoxic chemotherapy and even after neuropathy was established, acetyl carnitine significantly reduced its severity. Finally and most importantly, experiments in different tumor systems indicated that ALC does not interfere with the antitumor effects of chemotherapy. These findings lead the investigators to conclude that acetyl carnitine was effective in the prevention and treatment of chronic chemotherapy induced peripheral neurotoxicity and did not diminish the effectiveness of chemotherapy treatment in an experimental rat model.[255]

Considering the absence of any satisfactory treatment currently available for chemotherapy induced peripheral neuropathy (CIPN) in a clinical setting, these are important observations, opening up the possibility of using ALC to treat a wide range of patients who have undergone chemotherapy and developed sensory peripheral neuropathy.[256]

In human studies, acetyl carnitine treatment led to long-term symptomatic improvement in most patients without the need to discontinue neurotoxic chemotherapy drugs. Although in this study there was no control group, this agent, ALC, appeared to be an effective pathogenesis-based treatment for this type of chemotherapy related neuropathy. [257]

These findings are supported by another study which examined the affects of ALC on pathology of the nerves in chemotherapy patients. The investigators reported ALC treatment completely prevented the paclitaxel-evoked increase in the incidence of swollen and vacuolated C-fiber mitochondria, while having no effect on the paclitaxel-evoked changes in A-fiber mitochondria. These results suggest that the efficacy of prophylactic ALC supplementation against paclitaxel-evoked peripheral neuropathy pain may be related to a protective effect on C-fiber mitochondria. [258] C-fibers carry pain sensations from the periphery to the brain.

These results were corroborated in another investigation. The authors report that administration of inhibitors of second messengers implicated in models of other painful peripheral neuropathies (PKA, PKC, NO, Ca(2+), and caspase) had no effect, whereas both systemic and local administration of antioxidants (acetyl-L-carnitine, alpha-lipoic acid or vitamin C), all markedly inhibited oxaliplatin-induced hyperalgesia. [259]

In a clinical trial of ALC, at least one WHO grade improvement in the peripheral neuropathy severity score was shown in 73% of the patients receiving this nutrient. The treatment appeared to lack significant side affects with only a single case of insomnia related to ALC treatment reported. Given the safety and effectiveness of Acetyl-L-carnitine, it seems to be an effective and well-tolerated agent for the treatment of chemotherapy induced peripheral neuropathy. [260]

Taken together, these results suggest that ALC is not likely to interfere with chemotherapy, is safe and can prevent or even reverse the adverse neurotoxic affects of some chemotherapy treatments.

Antiviral Therapy

ALC has been given to HIV patients with symptomatic anti-retroviral neuropathy (ATN). In a number of clinical studies, ALC was administered either twice daily intramuscularly or as oral sachets or tablets. ALC was shown to significantly reduce pain as measured by a variety of validated pain ratings, and was generally safe and well tolerated.

The affects of ALC supplementation on the density of nerves fibers in patients suffering from ATN has likewise been studied. Using a measure of neuronal innervation in standardized skin biopsies of the affected area, cutaneous nerve density improved following the administration of ALC in subjects with symptomatic ATN. These patients all had previous evidence of reduced epidermal and dermal innervation, which was reversed by the ALC treatment. The histological changes induced by ALC were associated with clinical improvement, which was maintained over a 4-year period. Improvements were seen in both the structure and function of small sensory fibers, which were sustained over time while subjects remained on ALC supplements [262]

Acetyl carnitine, administered twice a day intramuscularly to HIV infected patients with symptomatic ATN significantly reduced weekly mean pain ratings on the VAS compared with placebo. Treatment with oral ALC improved symptoms for the patient group as a whole. Intramuscular and oral ALC therapy was considered generally safe and well tolerated. [263]

In this study, ALC was effective and well tolerated in symptomatic treatment of painful neuropathy associated with antiretroviral toxicity. Symptoms were reduced, even though no measurable effect was noted on neurophysiological parameters. [261]

Diabetes

There are many complications associated with diabetes. Both type one and type two diabetic patients are prone to develop neuropathy. The mechanisms that cause neuropathy in these two distinct types of diabetes appear to differ. Such differences in basic initiating factors and pathogenesis translate into differences in the functional and structural expressions of neuropathy in type 1 and type 2 diabetes. Type 1 neuropathy shows a more rapid progression with more severe functional and structural changes compared with type 2 diabetes. [266] Therefore discrepancies in underlying pathogenic mechanisms in the two types of diabetic neuropathy must be taken into account in the design of therapies. Treatment should target and be tailored to the key respective and distinct pathogenic mechanisms. Therapies that meet these criteria include replacement of acetyl-L-carnitine and replenishment of C-peptide in type 1 diabetic neuropathy. [266]

In one study, the mean serum-free L-carnitine levels in diabetic patients with complications was almost 25% lower than in diabetic patients with no complications. On the basis of the study results, carnitine supplementation in diabetic patients, especially in patients with diabetes complications, might be deemed useful. [265]

In an animal model of diabetic neuropathy, both prophylactic and therapeutic treatment with ALC did not affect the tail-flick latency in non-diabetic (control) mice. In diabetic mice however, pain behavior was significantly reduced. These findings indicate ALC does not affect general nociceptive neural transmission, but is specific to the pathology of the diabetic neuropathy. These results provide evidence of the prophylactic and therapeutic potential of ALC on the progression of diabetic neuropathy. [264]

Another study found that starting carnitine treatment in the early stages of diabetic neuropathy may be more effective in the treatment of sub-clinical neuropathy. The therapy should continue over the long term because a two-month trial of ALC treatment appeared to be insufficient to allow detection of electrophysiological improvement in cases where neurological deficits were clinically apparent. [267]

What these studies suggest is that ALC treatment may specifically reverse the pathology in nerves associated with diabetes and that long term administration is indicated and required to produce changes in nerve physiology.

Compressive Neuropathy

Acetyl-L-Carnitine also appears to be beneficial in compressive neuropathies like carpal tunnel syndrome and sciatica. In this study, it was found that physical decompression significantly improves the recovery rate of peripheral nerve as compared with treatments without decompression Also that acetyl-L-carnitine co-administered with decompression enhances clinical and histopathological recovery. [268]

Nicotinic Acetylcholine Receptors

Peripheral nerve injury has been shown to alter neural nicotinic receptors. [20] Furthermore substances that agonize the nicotinic acetylcholine receptors have been shown to produce direct pain inhibition. [269-274] This suggests that the acetylcholine nicotinic receptors may play a role in the pathogenosis and treatment of painful peripheral neuropathy.

A recent study suggests that the anti-nociceptive effects of acetyl carnitine may be enhanced by simultaneous stimulation of nicotinic acetylcholine receptors. [275] This suggests that niacin or one of its derivatives might be used along with acetyl carnitine to maximize the pain relieving affects of acetyl carnitine. Several studies suggest that Lemon Balm, Melissa Officinalis, has in addition to its effects on GABA metabolism (discussed elsewhere in the paper), a demonstrated agonistic effect on nicotinic acetylcholine receptors. [276-278] Suggesting that Lemon Balm might act synergistically with acetyl carnitine and enhance pain suppression in neuropathy patients.

Taurine

Taurine is an amino acid that exhibits several properties that make it another natural supplement that might benefit patients suffering from neuropathy. Taurine may play an important role in the prevention or reversal of complications of diabetes. It has several properties that are specific to the disease including the finding that taurine levels tend to be reduced in patients suffering from diabetes. In addition, taurine also influences ion channels and several neurotransmitters that are involved with the suppression of pain. Therefore taurine may have more widespread application in the management of other types of neuropathy unrelated to diabetes. Let's look at some of the biological properties of taurine. Taurine appears to modulate calcium levels through the calcium channels directly or indirectly through other mechanisms.

Previous research has shown that taurine has protective effects against glutamate-induced neuronal injury in cultured neurons. In this study the authors propose that the primary underlying mechanism of the neuroprotective function of taurine is due to its action in preventing or reducing glutamate-elevation of intracellular free calcium. [279]

In a review of the literature of the known actions of taurine, the authors highlight recent discoveries regarding taurine and calcium homeostasis in neurons. In general there is a consensus that taurine is a powerful agent in regulating and reducing the intracellular calcium levels in neurons. [280]

Two specific targets of taurine action are discussed: one is the Na(+)-Ca2+ exchanger, and the second is the metabotropic receptors mediating phospholipase-C. [280]

Taurine also acts like an inhibitory neurotransmitter. Either acting directly or through an agonism of glycine and GABA receptors. Glycine and GABA both are inhibitory neurotransmitters that are believed to suppress pain. Taken together these studies show that taurine might represent another important neurotransmitter or modulator in spinal cord Substantia Gelatinosa neurons. These spinal cord neurons are involved with pain gating at the spinal cord level. [281] This finding is corroborated by another investigation that indicates that taurine may act as a native ligand of the central glycine receptor serving to modulate neurotransmission in the immature hippocampus and also under certain conditions may activate GABA receptors directly. [282]

The role of taurine in diabetes and diabetic complications is well researched. It is hypothesize that glucose induced sorbitol accumulation in diabetes mellitus results in taurine depletion in peripheral nerve which may potentially impair nerve regeneration and precipitate neuronal hyper-excitability and pain. [161]

In fact the development of nerve conduction slowing in diabetes is accompanied by depletion of the amino acid taurine. Since taurine functions as an antioxidant, calcium modulator, and vasodilator, taurine depletion may provide a pathogenic link between the nerve metabolic, vascular and functional deficits complicating diabetes.[283]

The eventual development and clinical expression of diabetic neuropathy may result from progressive nerve fiber damage with blunted nerve regeneration and repair and may be complicated by nerve hyperexcitability resulting in pain. The naturally occurring amino acid taurine functions as an osmolyte, inhibitory neurotransmitter, and modulator of pain perception. It is also known to have neurotrophic actions. [161]

Oral taurine supplementation in diabetic patients was studied to determine if taurine might counteract oxidative stress and the nerve growth factor (NGF) deficits which have been observed in the diabetic peripheral nerve. [284] It was found that taurine administered orally, counteracts oxidative stress and the NGF deficit in early diabetic neuropathy. Additionally it was found the antioxidant effects of oral taurine in peripheral nerve are, at least in part, mediated through the ascorbate system of anti-oxidative defense. [284] Taurine also appears to ameliorate lipid-induced functional beta cell decompensation and insulin resistance in humans, possibly by reducing oxidative stress [285].

Given the diverse biological activity of taurine it appears that this amino acid has good potential for the management of diabetic complication like neuropathy and may also be important in correcting or ameliorating many of the metabolic defects associated with other forms of neuropathy.

GABA-ergic Nutrients

Introduction: GABA and neuropathy:

Inflammatory diseases and neuropathic insults are frequently accompanied by severe and debilitating pain, which can become chronic and often unresponsive to conventional analgesic treatment. [286] [287]

This process is believed to have a peripheral component located in the spinal cord and a central component in the higher centers of the nervous system.

The use of genetically engineered mice that lack specific GABA receptors has lead to a much greater understanding of the process of the development of chronic pain states and the role of the neurotransmitter GABA and it's receptors in the regulation of pain signaling in normal and neuropathic conditions. [286]

The data from these and other experiments suggests that neuropathic pain due to sensory nerve injury (similar to that occurring in neuropathy) is at least in part, the result of peripheral sensitization of neurons in the spinal cord. This sensitization leads to a long-lasting increase in synaptic plasticity in the spinal dorsal horn. A loss of GABA mediated synaptic inhibition in the spinal dorsal horn is believed to contribute significantly to the development and maintenance of this neuropathic pain pathology. There are many lines of evidence that suggest that the inhibitory spinal neurotransmitter, GABA, regulates this process through activation of specific spinal GABA receptors.

Long lasting down-regulation of GABA tone or diminished sensitivity to the inhibitory affects of GABA in the dorsal root ganglia neurons has been demonstrated in animal models of neuropathy. [287]

This data suggests that elevation of GABA and stimulation of GABA receptors offers a viable therapeutic target that may be able to compensate for the pathological changes (sensitization of neurons) that occur in the spinal cord of patients that suffer from neuropathy and other neuropathic pain states.

Specific GABA receptor activation not only diminished the nociceptive input to the brain from hyper-excitable spinal pain signaling neurons, but also may reduce the activity of brain regions related to the associative-emotional components of pain. Thus GABA appears to have a dual action on pain suppression both acting directly on sensitized spinal neurons and also centrally in those parts of the brain responsible for processing painful input from lower parts of the nervous system.

The accumulated basic science research has shown that potent pain suppression can be achieved by specifically targeting and activating GABA(A) receptors in the spinal cord. Furthermore functional magnetic resonance imaging (fMRI) in rats treated with GABA-ergic drugs correlates the clinical suppression of pain behavior with alterations in those parts of the brain that are known to relate to pain signal processing. These findings provide a rational basis for the use of GABA agonistic agents for the treatment of chronic refractory pain as an alternative to classical analgesic drugs. [286] Activation of GABA mediated inhibitory inputs from sensory neurons could be an attractive method to alleviate pain of neuropathic origin. This inhibitory GABA-ergic input may be enhanced   through somatosensory stimulation (TENs, Acupuncture, etc.), through elevation of GABA levels in the nervous system by providing oral GABA precursors, through herbal supplements known to modulate the enzymes that degrade GABA (GABA-T)   and synthesize GABA from Glutamate (GAD) or any combination of these methods

Lemon Balm:

Lemon balm (Melissa officinalis L.) is a lemon scented member of the mint family. A native to southern Europe, it is a perennial and grows to a height of about two feet. Historically Lemon balm was used in herbal medicine. It is believed to relax the nervous system producing sleep. It is used for epilepsy, nerve disorders, insomnia, fainting, hysteria, migraine headaches, hypochondria and vertigo.

Recently studies have demonstrated that lemon balm may influence GABA metabolism in the nervous system. The enzyme GABA transaminase (GABA-T) breaks down the neurotransmitter GABA and thus shuts off or at least diminishes the inhibitory action of this neurotransmitter. Substances that inhibit GABA-T will cause a net accumulation of GABA and indirectly potentiate the activity of GABA in the nervous system. In one study substances were tested for their influence on the GABA-T enzyme, the aqueous extract of Melissa Officinalis (Lemon Balm) exhibited the greatest inhibition of GABA-T activity of the substances tested.[288] Another study found extracts of Melissa Officinalis and Salvia Triloba had moderate activity at GABA receptors. [289] Yet another study revealed Lemon Balm elicited a significant dose-dependent reduction in both inhibitory and excitatory transmission, with a net depressant effect on neurotransmission[290]

Passiflora:

Passiflora Incarnata Linn. (Passion Flower) is used in several parts of the world as a traditional medicine for the management of anxiety, insomnia, epilepsy and morphine addiction. Recent research suggests that passiflora may exert its effects through modulation of the GABA receptors in the nervous system.

The potential anxiolytic effects of chrysin, a Passiflora extract, and its purported modulation of the benzodiazepine receptor or the GABA(A) receptor was studied in laboratory rats. [291] The results link chrysin to GABA receptors. The anxiolytic effect of chrysin, which was blocked by the injection of a benzodiazepine antagonist, could be linked to an activation of the GABA(A) receptor unit.[292] These findings were corroborated in another study which found the effects of specific receptor blockades on the inhibitory responses indicated that the herbal extract acts on gamma-amino butyric acid (GABA) receptors. [293] Extracts of passiflora may contain plant derived GABA. Passiflora extracts demonstrated antihypertensive effects on spontaneously hypertensive rats. This is likely explained by a GABA-induced antihypertensive activity. [294] Various extracts of Passiflora seem to exert their effects directly or indirectly through GABA mediated pathways. The results of this study show an anxiolytic profile for Passiflora Actinia. There are also indications of an involvement of GABA(A) system in this anxiolytic effect.[295] It seems that Pasipay, an extract of Passiflora, could be useful for treatment absence seizure. The effects of this Passiflora extract may be related to its modulation of GABA-nergic and opioid systems. More studies are needed in order to investigate its exact mechanism [296]

Valerian

Valerian root is used in the traditional medicine of many cultures as a mild sedative and to aid the induction of sleep. Valerian is a native plant both of Europe and North America. Valeriana Officinalis is the species most commonly used in Northern Europe. Valerian has few side effects

GABA-T

As discussed previously, substances that interfere with the action of GABA-T cause accumulation of GABA at the receptors and a net effect of dampening excitability of neurons. One study found effects of valerian inhalation reduced the activity of GABA-T and enhances GABA activity. [297] Another found Valerian extracts to have a direct action on the amygdaloid body of the brain and inhibit enzyme-induced breakdown of GABA in the brain resulting in sedation. [298]

GAD

Glutamate Decarboxylase or GAD is an enzyme that raises neural GABA through the conversion of Glutamate to GABA (discussed in detail later).

Extracts from Valeriana Officinalis (valerian) and Centella Asiatica (Gotu Kola) stimulated GAD activity by over 40% at a dose of 1 mg/mL. Thus potentially raising GABA levels in neural tissue and producing a dampening effect on nerve conduction. [288]

GABA

The valerian plant may actually contained plant derived GABA. Aqueous extracts of the valerian roots contain appreciable amounts of GABA [298]

GABA Receptor Activity

Evidence of a direct affect of valerian extract on the GABA receptor is accumulating. Several extracts of valerian enhanced the response to GABA at multiple types of recombinant GABA(A) receptors. [299] This finding was confirmed through studies employing specific GABA receptor blockade. The inhibitory responses provided evidence that the herbal extracts of valerian act on gamma-amino butyric acid (GABA) receptors [293] The mechanism of this action may be due to channel blocking activity of valerian extracts. Valerian was identified as a subunit specific modulator of GABA(A) receptors. [300] The results confirm that valerian extracts have effects on GABA(A) receptors, but can also interact at other pre-synaptic components of GABAergic neurons. [301] Yet another study found valerian extracts modulated GABA receptors in the rat brainstem. [302]

Gotu Kola

Gotu Kola also known as Centella Asiatica is an herb with the long history of use as a nerve tonic. In addition to its traditional use to treat nerve related conditions, Gotu Kola has beneficial effects in small blood vessels.

The herb was studied in a prospective, placebo-controlled, randomized trial to determine, whether total triterpenic fractions of Centella Asiatica is effective in improving the microcirculation in diabetic microangiopathy and neuropathy. [303] Gotu Kola through it's ability to diminish edema was found to be effective in the early stages of microangiopathy and has potential to help avoid progression to clinically significant stages.[303]

In another study, the effects of Gotu Kola on peripheral nerve pathology and regeneration were studied. The experimental group given Gotu Kola demonstrated axonal regeneration (larger caliber axons and greater numbers of myelinated axons) compared with controls, indicating that the axons grew at a faster rate. The authors of this study concluded that these findings indicated Centella ethanolic extract may be useful for accelerating repair of damaged neurons.[220]

Glutamine:

Several amino acids act as precursors to the neurotransmitters Glutamate and GABA. Glutamate causes neuron excitation whereas GABA has an inhibitory influence on neural transmission. [304-308] The pathways that lead to the synthesis of Glutamate and GABA are intimately interrelated. The enzyme GABA transaminase serves to break down GABA thus attenuating its neuro-inhibitory effects on nerve transmission. Stated more simply, GABA inhibits nerve activity, GABA-T reduces nerve GABA thus ultimately causing a net increase in nerve excitability.

The enzyme Glutamate Decarboxylase (GAD) degrades the excitatory amino acid, Glutamate, thus causing a net suppression of nerve transmission. More simply put, Glutamate excites or facilitates nerve activity, GAD by degrading Glutamate attenuates Glutamate's neuro-excitatory activity. [304-308]

The relationship between neural excitation through Glutamate and neural inhibition through GABA is even more complex due to the fact that the degradation of Glutamate via the GAD enzyme actually produces GABA. So much so that is it believed that the main source of GABA in the nervous system is a direct result of GAD activity on Glutamate.

Further research suggests that the GAD enzyme is the rate limiting step in the elevation of GABA levels in the brain. Under normal circumstances, this enzyme is unsaturated.

Again in an attempt to simplify a highly complex metabolic concept, we can think that there is usually a shortage of Glutamate available to the GAD enzyme. Thus, GABA levels are lower than they would be, if more Glutamate were available to the GAD enzyme to convert into GABA.

This provides a pathway with which to increase GABA levels in the nervous system for the purpose of suppression or inhibition of neural activity. Increased Glutamate levels may saturate the GAD enzyme and ultimately increase GABA levels.

Some authors question if oral GABA leads to increased levels of this neurotransmitter in the brain. There is controversy about the ability of orally administered GABA to cross the blood brain barrier.

The amino acid glutamine, however, can and does cross the blood brain barrier. It serves as a precursor to Glutamate which is enzymatically converted by GAD into GABA in the nervous system.

Several studies have demonstrated that oral supplementation with Glutamine leads to increased levels of GABA in nervous tissue, presumable through the Glutamate-GAD-GABA shunt just described.

This makes oral glutamine supplementation an ideal treatment designed to nutritionally raise GABA levels in nervous tissue. It would appear that glutamine therapy may raise neural GABA and thus reduce hyper-excitability in neurons as is found in patients suffering from peripheral neuropathies.

A number of clinical trials have demonstrated the ability of Glutamine to ameliorate the signs and symptoms of peripheral neuropathies associated with specific types of chemotherapy. [309-318] The effectiveness of glutamine in the treatment of these neuropathies may be explained, at least in part, by the elevation of GABA believed to occur following oral supplementation of this substance. Elevation of the inhibitory amino GABA may not be the only mechanism by which glutamine suppresses symptoms of neuropathy. A few studies suggest that glutamine may also stimulate nerve growth factor and or growth hormone secretion. [319] Finally, glutamine supplementation has a very good safety provide with few, if any, reported adverse effects. [320-322]

Vinpocetine:

Vinpocetine (pronounced vin-poe-ce-teen) is a synthetic compound derived from vincamine, a substance found naturally in the leaves of the lesser periwinkle plant (Vinca minor). Vinpocetine was developed in the late 1960s.

Vinpocetine is available as a prescription drug in Europe and Japan. In the United States and Canada, it is sold in health food stores and online as a dietary supplement.

Vinpocetine exhibits a diverse pharmacological profile that includes action at several ion channels, principally "generic" populations of sodium channels that give rise to tetrodotoxin-sensitive action potentials. A number of cell types are known to express tetrodotoxin-resistant (TTXr) sodium related action potentials. The molecular basis of TTX resistant impulse conductance has remained elusive until recently. One TTX resistant channel, named NaV1.8, is of particular interest because of its prominent and selective expression in peripheral afferent nerves. [323] Afferent nerves as we have discussed throughout this book are related to sensory symptoms of neuropathy, like pain and numbness. The present available data demonstrate that vinpocetine is capable of blocking NaV1.8 sodium channel activity and suggest a potential additional utility in various sensory abnormalities arising from abnormal peripheral nerve activity. [323]

Let's look at some other studies of vinpocetine and how this plant derivative might be used as part of a comprehensive plan to treat neuropathy.

In the medical management of the symptoms of neuropathy, anti epileptic drugs are the mainstay of treatment. There is evidence that vinpocetine exerts anticonvulsive or antiepileptic activity.

Vinpocetine at doses from 2 to 10 mg/kg inhibits the tonic-clonic convulsions induced by PTZ, a convulsing agent. Vinpocetine given to animals by injection at a dose of 2 mg/kg, 4 hours prior to PTZ administration, completely prevented the characteristic EEG changes (electroencephalogram) commonly produced by PTZ. Vinpocetine also abolished the PTZ-induced changes in the amplitude and latency of the later waves of the Brainstem Auditory Evoke Potentials in response to pure tone burst stimuli [324]

These results show the antiepileptic potential of vinpocetine and indicate the capability of vinpocetine to prevent the changes in the BAEP waves associated with the hearing loss observed during generalized epilepsy. [324]

In another study designed to investigate the underlying mechanism of the antiepileptic activity of vinpocetine, vinpocetine was found to abolished the increase in intracellular sodium influx that is induced by the convulsive agent M 4-AP similarly to the Na(+) channel blocker, tetrodotoxin. [325] It was concluded that the inhibitory effect of vinpocetine on pre-synaptic voltage-sensitive sodium channels might at least partially explain the anticonvulsant action of this natural substance. [325]

Vinpocetine was compared directly with other known anticonvulsive agents (AEDs). The study found that all of the antiepileptic substances likely exerted their affect by blocking the pre-synaptic sodium currents

120

more so than through effects on the pre-synaptic calcium channel mediated responses. Of the Anti-Epileptic Drugs studied, vinpocetine displayed the highest potency at the lowest dose. [326]

Most available AEDs require high doses to control seizures and are frequently accompanied by adverse secondary effects. The higher efficacy and potency of vinpocetine acting to reduce the permeability of pre-synaptic ionic channels and thus controlling the release of the excitatory neurotransmitters in the brain, leads the authors to conclude that vinpocetine may be advantageous in the treatment of epilepsy. [326]

Another study corroborates the above findings suggesting that the main mechanism involved in the neuroprotective action of vinpocetine in the CNS is unlikely to be due to a direct inhibition of Ca2+ channels, but rather the inhibition of pre-synaptic Na+ channel-activation unchained responses. [327]

These and other studies suggest that vinpocetine may be a promising alternative for the treatment of epilepsy.[328]

One additional study that is worth mentioning deals with vinpocetine's affect on the NMDA receptor. This central nervous system receptor is believed to play a role in chronic pain processing in the spinal cord and brain. This is particularly intriguing for the treatment of neuropathic pain in which it is generally believed that there is an opioid-insensitive component that can be blocked by NMDA receptor antagonists. However, in order to obtain complete analgesia, a combination of an NMDA receptor antagonist and an opioid receptor agonist is needed.

Vinpocetine demonstrated dose dependant antagonism of the NMDA receptor. Bifemelane, indeloxazine and vinpocetine suppressed the maximum response of NMDA and Glycine [329] These results suggest that the inhibition of NMDA channels by vinpocetine shows a similarity to the action of Zn2+ which closes the gate of the NMDA channel [329] These results suggest that vinpocetine may have anti-nociceptive potential in cases of peripheral neuropathy.

Section III: External Topical Treatments and Miscellaneous Neuropathy Therapies.

Herbal Liniment

There are several topical ointments or rubs that we have found to be effective for the treatment of our patient's neuropathy symptoms.

The first is a cream made from hot peppers. This is known as capsaicin cream. Capsaicin is the chemical that makes pepper plants taste hot. It also has a unique ability to drain the neurotransmitter called Substance P (named Substance P because it is associated with the transmission of pain impulses through the nervous system to the brain) from nerve endings in the skin. With repetitive use, Capsaicin cream will deplete enough Substance P from the nerves that they will no longer be able to transmit pain signals.

Diabetic painful neuropathy: current and future treatment options. Chong MS, Hester J. **Drugs. 2007;67(4):569-85.**

Diabetic painful neuropathy (DPN) is one of the most common causes of neuropathic pain. The management of DPN consists of excluding other causes of painful peripheral neuropathy, maximizing diabetic control and using medications to alleviate pain. The precise relationship between glycaemic control and the development and severity of DPN remains controversial. In this context, drugs such as aldose reductase inhibitors, ACE inhibitors, lipid-lowering agents and alpha-lipoic acid (thioctic acid) may have a useful role to play. together with capsaicin are effective for alleviating DPN. There is also good evidence that the serotonin-noradrenaline reuptake inhibitor antidepressant drugs venlafaxine and duloxetine are effective for treating DPN. *Capsaicin has the best evidence base of all the topical agents, but local anesthetics patches may also have a useful therapeutic role.*

The second is a Chinese Liniment call Green Willow Liniment. Green Willow is based on a formula that first appeared in texts of Traditional Chinese Medicine published thousands of years ago. The ingredients for Green Willow Ligament include:

1. Carthamus, 2. Aconitum, 3. Angelica Sinensis, 4.Persica, 5. Zingiber, 6. Glycyrrhiza, 7.Rheus, 8.Pyritum, 9. Strychnos, 10. Cinnamomum, 11.Auklandia, 12. Myrrha.

There are a few studies that look at the anti-nociceptive and anti-inflammatory properties of the ingredients contained in Green Willow Formula. Anti-nociception is the ability to block pain, anti-inflammatory means the ability to block pain and inflammation in damaged tissues. Most of the ingredients in this formula have pain and inflammation regulating properties. Let's look at a few of the studies and their findings for the individual components of the Green Willow Formula.

1.Antiinflammatory and analgesic effects of Carthamus lanatus aerial parts. Bocheva A **Fitoterapia. 2003 Sep;74(6):559-63.**

Carthamus lanatus aerial parts given by oral route at a dose of 2 mg/kg showed significant antiinflammatory activities in rats. The alcohol extract possesses a significant analgesic activity.

2. Bulleyaconitine A isolated from aconitum plant displays long-acting local anesthetic properties in vitro and in vivo. Wang CF Anesthesiology. **2007 Jul;107(1):82-90.**

BACKGROUND: Bulleyaconitine A (BLA) is an active ingredient of Aconitum bulleyanum plants. BLA has been approved for the treatment of chronic pain and rheumatoid arthritis in China. Single injections of BLA (0.2 ml at 0.375 mm) into the rat sciatic notch blocked sensory and motor functions of the sciatic nerve. When BLA at 0.375 mm was coinjected with 2% lidocaine (approximately 80 mm) or epinephrine (1:100,000) to reduce drug absorption by the bloodstream, the sensory and motor functions of the sciatic nerve remained fully blocked for approximately 4 h and regressed completely after approximately 7 h, with minimal systemic effects.

3. The aqueous extract of a popular herbal nutrient supplement, Angelica sinensis, protects mice against lethal endotoxemia and sepsis. Wang H
J Nutr. 2006 Feb;136(2):360-5.

The high mortality from sepsis is in part mediated by bacterial endotoxin, which stimulates macrophages/monocytes to sequentially release early (e.g., tumor necrosis factor, interleukin-1, and interferon-gamma) and late [e.g., high mobility group box 1 protein (HMGB1)] proinflammatory cytokines.

Our discovery of HMGB1 as a late mediator of lethal systemic inflammation has initiated a new field of investigation for the development of experimental therapeutics. Prophylactic administration of an aqueous extract of A. sinensis significantly attenuated systemic HMGB1 accumulation in vivo, and conferred a dose-dependent protection against lethal endotoxemia. Furthermore, delayed administration of A. sinensis extract beginning 24 h after CLP attenuated systemic HMGB1 accumulation, and significantly rescued mice from lethal sepsis. Taken together, these data suggest that A. sinensis contains water-soluble components that exert protective effects against lethal endotoxemia and experimental sepsis in part by attenuating systemic accumulation of a late proinflammatory cytokine, HMGB1.

4. Antinociceptive effect of Phlomis olivieri Benth., Phlomis anisodonta Boiss. and Phlomis persica Boiss. total extracts. Sarkhail P, Abdollahi
Pharmacol Res. 2003 Sep;48(3):263-6.

In this study total extracts of Phlomis Persica were tested for their antinociceptive effects. total extracts P. persica, at doses of 150, 150 and 100 mg kg(-1), showed significant antinociceptive effects. Results also showed that antinociceptive properties of P. olivieri, P. anisodonta and P. persica with ED(50) values of 88.21, 123.62 and 59.24 mg kg(-1), respectively, are comparable to that of indomethacin (5 mgkg(-1)).

5. Analgesic, antiinflammatory and hypoglycaemic effects of ethanol extract of Zingiber officinale (Roscoe) rhizomes (Zingiberaceae**) in mice and rats.
Ojewole JA. Phytother** Res. 2006 Sep;20(9):764-72.

The present study was undertaken to investigate the analgesic, anti-inflammatory and hypoglycaemic effects of Zingiber officinale dried rhizomes ethanol extract The findings of this experimental animal study indicate that Zingiber officinale rhizomes ethanol extract possesses analgesic and anti-inflammatory properties; and thus lend pharmacological support to folkloric, ethnomedical uses of ginger in the treatment and/or management of painful, arthritic inflammatory conditions.

6.Anti-inflammatory effect of roasted licorice extracts on lipopolysaccharide-induced inflammatory responses in murine macrophages. Kim JK,
Biochem Biophys Res Commun. 2006 Jul 7;345(3):1215-23

Licorice, the roots of Glycyrrhiza inflata, is used by practitioners of alternative medicine to treat individuals with gastric or duodenal ulcers, bronchitis, cough, arthritis, adrenal insufficiency, and allergies. In the murine model, we found that in vivo exposure to Licorice root extract an increase in the survival rate, reduced plasma levels of TNF-alpha and IL-6, and increased IL-10 production in LPS-treated mice. Collectively, thesedata suggest that the use of licorice root extract may be a useful therapeutic approach to various inflammatory diseases

6. Analgesic and anti-inflammatory properties of brucine and brucine N-oxide extracted from seeds of Strychnos nux-vomica.
Yin W. **J Ethnopharmacol. 2003 Oct;88(2-3):205-14.**

To further understand the purpose of the traditional processing method of the seeds of Strychnos nux-vomica L. (Loganiaceae) as well as analgesic and anti-inflammatory activities of brucine and brucine N-oxide extracted from this medicinal plant, various pain and inflammatory models were employed in the present study to investigate their pharmacological profiles.

These results suggest that central and peripheral mechanism are involved in the pain modulation and anti-inflammation effects of brucine and brucine N-oxide, biochemical mechanisms of brucine and brucine N-oxide are different even though they are similar in chemical structure.

7. Myrrh: medical marvel or myth of the Magi?
Nomicos EY. **Holist Nurs Pract. 2007 Nov-Dec;21(6):308-23**

Since antiquity, the genus Commiphora is composed of more than 200 species, and has been exploited as a natural drug to treat pain, skin infections, inflammatory conditions, diarrhea, and periodontal diseases.

So these studies suggest that the individual components of the Green Willow formula have varying degrees of both anti-inflammatory and anti-nociception (pain killing) properties that work through a wide range of pathways in the nervous system.

Our patients uniformly feel that this liniment is very helpful in the symptomatic relief of their peripheral neuropathy.

St. Johns Wort and Kava [330, 331], [332-337]

Two additional substances have potential for the treatment of painful neuropathy. St. John's Wort (SJW) is well known as a natural anti-depressant. Kava is a well known anti-anxiety herb

Both SJW and Kava appear to modulate sodium ion channels in the nervous system. If you remember our discussion of these channels earlier in this paper, sodium channels are responsible for fine tuning the nerve axon, making it either more or less likely to send a painful signal to the brain. These two herbs, when used topically, may influence the ion channels in the peripheral nerves and reduce their excitability, something that would appear to benefit patients suffering from peripheral neuropathy. We mix these herbs into our ultrasound gel and use them as a topical lotion on the feet of our neuropathy patients.

Caution is urged with the consumption of these herbs however. St Johns Wort affects a particular enzyme in the liver. This enzyme is responsible for the metabolism of many drugs. Since most neuropathy patients are taking multiple medications, the internal use of SJW should be used with extreme caution. Consumption of Kava has been associated with rare reports of extreme liver toxicity. Internal consumption of these herbs should be done only under medical supervision.

Kampo Medicine [338, 339]

Kampo Medicine is the Japanese reformulation of Traditional Chinese Herbal Medicine Remedies. Kampo herbal remedies are subject to strict pharmaceuticals processing and manufacturing procedures. Many of the Kampo remedies have been subjected to rigorous controlled clinical trials to test for their effectiveness. Two particular Kampo formulations, Gosha-jinki-gan and Shakuyakukanzoto, have shown potential for the treatment of diabetes and diabetic neuropathy.

Section IV: Results and future direction.

Results: There is Help ~ Johns Hopkins

Our Results

We presented a consecutive case series of the results our patient obtained in our office, using the methods and techniques we developed and described in this book. The average patient, regardless of the type or duration of their neuropathy, improved 50- 70% after six weeks of therapy. Below is the summary of the research we conducted on our techniques and presented at a Post Graduate Conference at Turner's Hall at Johns Hopkins Medical School in Baltimore, Maryland in June 2007.

Peripheral Neuropathy an Integrative Medical Approach: Results of a Consecutive Case Series.

Background Data:

Peripheral neuropathy is a common neurological condition. There are several well-documented causes of peripheral neuropathy, including complications from diabetes and chemotherapy, vitamin deficit, and exposure to neurotoxins. Many cases of peripheral neuropathy are classified as idiopathic or hereditary. Currently, medical therapy seeks to control the pain, burning, and numbness of peripheral neuropathy through medications. GABA-nergic medications are the mainstay of medical management. Often medical management is inadequate. A consecutive case series of 11 patients suffering from peripheral neuropathy of various etiologies, treated with an integrative medical approach is presented.

Interventions & Outcomes:

Physical treatments included needle acupuncture, low level laser therapy, and ultrasound. Nutritional supplementation included L-Acetyl Carnitine, Gotu Kola, and Deer Antler Velvet. Short term reduction in symptoms as measured by VAS was reported in all 11 patients. Long term improvement in pain and functional ability ranged from 50-70%. Two patients reported complete and long lasting amelioration of their symptoms.

Conclusions:

This consecutive case series of patients suffering from peripheral neuropathy of various etiologies reported both short and long term improvements in their symptomatology and ability to function. This integrative medical approach to neuropathy should be more formally studied with better and more objective outcome measures. If similar results can be replicated in a controlled investigation, an integrative approach to neuropathy may play a more prominent role in the treatment of this common neurological condition.

Case Study: Jill

Case studies when presented from the doctor's or researcher's perspective seldom describe results from a patient's point of view. So often medical researchers determine the effectiveness of a treatment based on a statistical analysis of data from a group of patients. In clinical practice a doctor treating neuropathy might be delighted when his patient has improved test results. However an improvement of 50-70% measured by a researcher or a 15% improvement on nerve conduction studies measured in clinical practice may or may not be meaningful to the patient.

It is the patient's perspective of how the treatment improved his or her quality of life that really matters. After all, it is the patient, not the statistic or the test that suffers from neuropathy.

On the following page is an account from one of our patients. It is unedited, except for removing her last name. It is the patient's perspective and the patient's results that truly matter. You can read Jill's response to our treatment in her own words on the next page.

Although it is only one patient's account, it is a truly representative case study from the patient's perspective. I could have included hundreds more just like Jill's.

In her own words.....

J M   EY
_____

From: "J M   Y" <jn   ey1@msn.com>
To: <info@neuropathy.org>
Sent: Wednesday, November 28, 2007 8:50 PM
Subject: Recommendation for a Dr.

Dear Sirs;

I became disabled with neuropathy in January of 2005. After many visits to neurologists, orthopedic doctors, pain doctors, psychologists and god only knows who else, I have finally been diagnosed with idiopathic peripheral neuropathy. It has been documented that my condition is severe, chronic and progressive and very painful. This condition impairs my ability to ambulate and the strong medication (900mg x 3 times a day of neurontin and nortriptilyn nightly) that I "took" daily significantly impaired my ability to concentrate, resulting in me being unemployable. I was told by my doctors that I was headed for a wheelchair and that I simply had to "put up with it" and resign myself to the fact that I was never going to get back to normal again. How do these doctors expect for you to even want to get better if this is their diagnosis and they are not even going to do anything to help you get better.

In March of 2007, there was a flyer in my local newspaper for Dr. George Kukurin. Dr. Kukurin is a chiropractic physician, board certified in neurology and also practices acupuncture. I knew that the medication that I was taking was hurting my liver (this has been confirmed) and I was not improving at all and I had previously had wonderful results with chiropractic physicians so I made an appointment with Dr. Kukurin. He truly was and is an answer to prayer. My first visit was on March 20th, 2007. I walked into his office that day with shuffling feet, little tiny steps and holding on to the walls. He told me that day that he would have me up and walking and back to normal by the end of September. I didn't say it, but I thought, "sure big guy, sure!!" Initially I was seeing him about every other day. Every time I would see Dr. Kukurin, he did 5 - that's right - 5 different treatments on me. He would start out by relaxing the muscles in the relaxation seat (the one that rolls up and down your spine) and follow that up with a laser, ultrasound, TENS, acupuncture and finally a chiropractic adjustment for my lower back. I would usually spend an easy hour in his office every other day. He also recommended that I take some supplements that included Antler Velvet, Acetyl L-Carnitine and a Ginseng Root formula. It wasn't long before I was only seeing him once a week and then, after about 5 months, I was getting around with relative ease. No more shuffling or hanging onto walls, I was leading a somewhat normal life. This man has brought me back to as close to normality as anyone could hope for. I had been bedridden for four months and I could not stand, walk or sit for more than about one hour at a time. Now, I am not taking the Neurontin at all. I can go for outings and can last for 3 to 4 hours instead of just an hour. I can enjoy my grandchildren and play with them at the park instead of just watching them play. I still have my limitations - I'm not totally normal but I will certainly take what Dr. Kukurin has given me in place of just taking medication that I knew was hurting me and would shorten my life-span with no hope for improvement. Will this work for everybody? I don't know, but I had to share this wonderful man with any and everybody that will listen. I won't close any doors on the opportunity to get better and relief from pain - I hope you won't either.

Dr. Kukurin has offices both on the East coast (Pittsburgh) and in the west (Arizona). **If you have neuropathy,** I would make every effort to find him. You can find his web-site at www.alt-compmed.com. His offices in Arizona are in Peoria and Goodyear. You can reach him in Arizona at 623-972-8400.

Dr. KuKurin listens to your complaints and then educates you about his alternative treatment plans. He and his staff are compassionate, understanding and sympathetic, always willing to help

Summary:

As we have discussed neuropathy is a debilitating condition with no known effective treatment. Many diseases and even the medical treatment of other diseases can cause neuropathy. While the standard medical approach is most often inadequate, alternative medicine and integrative neurology offer great hope to patients suffering with this dreaded condition. The techniques discussed in this manuscript have proven successful for a great many of our patients. Some of the results have been astounding. With the continued research into the neurobiology of neuropathy and the explosion of research into alternative medicine, we anticipate that the results our patients experience will likewise improve. It is my hope that neuropathy patients and their families will use the information in this manuscript not only to find hope and help with the pain of neuropathy, but also spread the word to the thousands of patients, currently suffering from neuropathy, who have never heard of integrative neurology.

From small acorns grow the mighty oak

Dedication:

In Loving Memory of My grandmother

Marie (Mary) F Novak

&

In Memory of My Friend and Colleague

Dr. Juli Rissman-Lewandosky

The star that burns twice as bright, burns half as long

Appendix:

Neuropathy Intake form & Outcomes Instrument

Name:

Date:

Right Sole

1. Toes

2. Mid-Foot

3. Heel

[ ] burning [ ] numb [ ] pin/needles [ ] shooting pain [ ]
ache

[ ] burning [ ] numb [ ] pin needles [ ] shooting pain [ ]
ache

[ ] burning [ ] numb [ ] pin needles [ ] shooting pain [ ]
ache

3. Heel

2. Mid-Foot

1. Toes

Left Sole

1. How bad is your pain right now?　　/ 10

2. How bad is your average or typical pain?　　/10

3. What is your pain level when you are feeling your best?　　/10

4. What is your pain level when you are at your

What % of the time are you at your best ?　　24%　50%　75%

What % of the time are you at your worst?　24%　50%　75%

Signed:　　　　　　　　Dated:

Table One: Nutrients that may have application in the treatment of neuropathy.

Nutrient	Neuropathy Clinical Trial	Na+ Channel Blocker	Ca+ Channel Blocker	PKC Inhibitor	Aldose Reductase Inhibitor	NGF	GABA
Acetyl Carnitine	XXX					X	
Taurine	XX		X			X	XX
Lipoic Acid	XXX						
Ginko				X			
Vitamin E				X			
Ginseng			XX	X			X
Glutamine	XXX					X	XX
Ascorbic Acid					X		
Quercetine					X		
Thiamine	X						
Gotu Kola						X	
Vanadyl Sulfate						X	
Elk Velvet						X	
Vinpocetine		X	X				X
Lemon Balm		X					X
Passiflora		X					X
Valerian		X					X

Acknowledgements:

The author would like to thank and acknowledge Ms. Cheryl Unterschutz
for creation of most of the graphics used in this text

.

The author would like to also thank and acknowledge Ms. Jill Nunn
for assisting in editing and manuscript preparation.

References

1.   Attal, N. and D. Bouhassira, *Mechanisms of pain in peripheral neuropathy.* Acta Neurol Scand Suppl, 1999. 173: p. 12-24; discussion 48-52.

2.   Kajdasz, D.K., et al., *Duloxetine for the management of diabetic peripheral neuropathic pain: evidence-based findings from post hoc analysis of three multicenter, randomized, double-blind, placebo-controlled, parallel-group studies.* Clin Ther, 2007. 29 Suppl: p. 2536-46.

3.   Wernicke, J.F., et al., *Safety and tolerability of duloxetine treatment of diabetic peripheral neuropathic pain between patients with and without cardiovascular conditions.* J Diabetes Complications, 2008.

4.   Sonnett, T.E., S.M. Setter, and R.K. Campbell, *Pregabalin for the treatment of painful neuropathy.* Expert Rev Neurother, 2006. 6(11): p. 1629-35.

5.   Vinik, A., *CLINICAL REVIEW: Use of antiepileptic drugs in the treatment of chronic painful diabetic neuropathy.* J Clin Endocrinol Metab, 2005. 90(8): p. 4936-45.

6.   Thorsteinsson, G., *Chronic pain: use of TENS in the elderly.* Geriatrics, 1987. 42(12): p. 75-7, 81-2.

7.   Klassen, A., et al., *High-tone external muscle stimulation in end-stage renal disease: effects on symptomatic diabetic and uremic peripheral neuropathy.* J Ren Nutr, 2008. 18(1): p. 46-51.

8.   Faingold, C.L., *Electrical stimulation therapies for CNS disorders and pain are mediated by competition between different neuronal networks in the brain.* Med Hypotheses, 2008.

9.   Mathis, F., et al., *[Peripheral neuropathic pain relieved by somatosensory rehabilitation].* Rev Med Suisse, 2007. 3(135): p. 2745-8.

10.   Ro, L.S. and K.H. Chang, *Neuropathic pain: mechanisms and treatments.* Chang Gung Med J, 2005. 28(9): p. 597-605.

11.   Hamza, M.A., et al., *Percutaneous electrical nerve stimulation: a novel analgesic therapy for diabetic neuropathic pain.* Diabetes Care, 2000. 23(3): p. 365-70.

12.   Rao, V.R., S.L. Wolf, and M.R. Gersh, *Examination of electrode placements and stimulating parameters in treating chronic pain with conventional transcutaneous electrical nerve stimulation (TENS).* Pain, 1981. 11(1): p. 37-47.

13.   Pinto, V., V.A. Derkach, and B.V. Safronov, *Role of TTX-sensitive and TTX-resistant sodium channels in Adelta- and C-fiber conduction and synaptic transmission.* J Neurophysiol, 2008. 99(2): p. 617-28.

14.   Aurilio, C., et al., *Ionic channels and neuropathic pain: physiopathology and applications.* J Cell Physiol, 2008. 215(1): p. 8-14.

15.   Valia-Vera, J.C., et al., *[Ziconotide: an innovative alternative for intense chronic neuropathic pain].* Rev Neurol, 2007. 45(11): p. 665-9.

16.   Jagodic, M.M., et al., *Cell-specific alterations of T-type calcium current in painful diabetic neuropathy enhance excitability of sensory neurons.* J Neurosci, 2007. 27(12): p. 3305-16.

17.   Honda, K., et al., *Contribution of Ca(2+) -dependent protein kinase C in the spinal cord to the development of mechanical allodynia in diabetic mice.* Biol Pharm Bull, 2007. 30(5): p. 990-3.

18.   Saegusa, H., et al., *Suppression of inflammatory and neuropathic pain symptoms in mice lacking the N-type Ca2+ channel.* EMBO J, 2001. 20(10): p. 2349-56.

19.   Landmark, C.J., *Targets for antiepileptic drugs in the synapse.* Med Sci Monit, 2007. 13(1): p. RA1-7.

20.   Young, T., et al., *Peripheral nerve injury alters spinal nicotinic acetylcholine receptor pharmacology.* Eur J Pharmacol, 2008. 590(1-3): p. 163-9.

21.   Slavin, K., *Peripheral nerve stimulation for neuropathic pain.* Neurotherapeutics 2008. 5(1): p. 100-6.

22.   Forst, T., et al., *Impact of low frequency transcutaneous electrical nerve stimulation on symptomatic diabetic neuropathy using the new Salutaris device.* Diabetes Nutr Metab, 2004. 17(3): p. 163-8.

23.   Kumar, D. and H.J. Marshall, *Diabetic peripheral neuropathy: amelioration of pain with transcutaneous electrostimulation.* Diabetes Care, 1997. 20(11): p. 1702-5.

24.   Inoue, T., et al., *Long-lasting effect of transcutaneous electrical nerve stimulation on the thermal hyperalgesia in the rat model of peripheral neuropathy.* J Neurol Sci, 2003. 211(1-2): p. 43-7.

25.   Julka, I.S., M. Alvaro, and D. Kumar, *Beneficial effects of electrical stimulation on neuropathic symptoms in diabetes patients.* J Foot Ankle Surg, 1998. 37(3): p. 191-4.

26.   Leem, J.W., E.S. Park, and K.S. Paik, *Electrophysiological evidence for the antinociceptive effect of transcutaneous electrical stimulation on mechanically evoked responsiveness of dorsal horn neurons in neuropathic rats.* Neurosci Lett, 1995. 192(3): p. 197-200.

27.   Maeda, Y., et al., *Release of GABA and activation of GABA(A) in the spinal cord mediates the effects of TENS in rats.* Brain Res, 2007. 1136(1): p. 43-50.

28.   Lee Son, G.M., J.S. Blouin, and J.T. Inglis, *Short duration galvanic vestibular stimulation evokes prolonged balance responses.* J Appl Physiol, 2008.

29.   Deutschlander, A., et al., *Unilateral vestibular failure suppresses cortical visual motion processing.* Brain, 2008. 131(Pt 4): p. 1025-34.

30.   Dieterich, M. and T. Brandt, *Functional brain imaging of peripheral and central vestibular disorders.* Brain, 2008.

31.   Gong, X., et al., *[Functional localization of vestibular cerebral representations in human using functional magnetic resonance imaging].* Zhonghua Er Bi Yan Hou Tou Jing Wai Ke Za Zhi, 2006. 41(10): p. 731-5.

32.   Schwoebel, J., et al., *Pain and the body schema: evidence for peripheral effects on mental representations of movement.* Brain, 2001. 124(Pt 10): p. 2098-104.

33.   Meng, H., et al., *Vestibular signals in primate thalamus: properties and origins.* J Neurosci, 2007. 27(50): p. 13590-602.

34.   Stephan, T., et al., *Functional MRI of galvanic vestibular stimulation with alternating currents at different frequencies.* Neuroimage, 2005. 26(3): p. 721-32.

35.   Kennedy, P., *Vestibulospinal influences on lower limb motoneurons.* Can J Physiol Pharmacol, 2004. 82(8-9): p. 675-81.

36.   Been, G., et al., *The use of tDCS and CVS as methods of non-invasive brain stimulation.* Brain Res Rev, 2007. 56(2): p. 346-61.

37.   Ramachandran, V.S., et al., *Rapid relief of thalamic pain syndrome induced by vestibular caloric stimulation.* Neurocase, 2007. 13(3): p. 185-8.

38.   Ramachandran, V.S., P.D. McGeoch, and L. Williams, *Can vestibular caloric stimulation be used to treat Dejerine-Roussy Syndrome?* Med Hypotheses, 2007. 69(3): p. 486-8.

39.   McGeoch, P.D., et al., *Behavioural evidence for vestibular stimulation as a treatment for central post-stroke pain.* J Neurol Neurosurg Psychiatry, 2008.

40.   McGeoch, P.D. and V.S. Ramachandran, *Vestibular stimulation can relieve central pain of spinal origin.* Spinal Cord, 2008.

41.   Cuccurazzu, B., et al., *A monosynaptic pathway links the vestibular nuclei and masseter muscle motoneurons in rats.* Exp Brain Res, 2007. 176(4): p. 665-71.

42.   Bohotin, C., et al., *Vagus nerve stimulation attenuates heat- and formalin-induced pain in rats.* Neurosci Lett, 2003. 351(2): p. 79-82.

43.   Hord, E.D., et al., *The effect of vagus nerve stimulation on migraines.* J Pain, 2003. 4(9): p. 530-4.

44.   Kirchner, A., et al., *Vagus nerve stimulation suppresses pain but has limited effects on neurogenic inflammation in humans.* Eur J Pain, 2006. 10(5): p. 449-55.

45.   Mauskop, A., *Vagus nerve stimulation relieves chronic refractory migraine and cluster headaches.* Cephalalgia, 2005. 25(2): p. 82-6.

46.   Multon, S. and J. Schoenen, *Pain control by vagus nerve stimulation: from animal to man...and back.* Acta Neurol Belg, 2005. 105(2): p. 62-7.

47.   Ansari, S., K. Chaudhri, and K.A. Al Moutaery, *Vagus nerve stimulation: indications and limitations.* Acta Neurochir Suppl, 2007. 97(Pt 2): p. 281-6.

48.   Kirchner, A., et al., *Left vagus nerve stimulation suppresses experimentally induced pain.* Neurology, 2000. 55(8): p. 1167-71.

49.   Fallgatter, A.J., et al., *Far field potentials from the brain stem after transcutaneous vagus nerve stimulation.* J Neural Transm, 2003. 110(12): p. 1437-43.

50.   Kraus, T., *BOLD fMRI deactivation of limbic and temporal brain structures and mood enhancing effect by transcutaneous vagus nerve stimulation.* J Neural Transm 114(11):1485-93, 2007. 114(11): p. 1485-93.

51.   Sallam, H., et al., *Transcutaneous electrical nerve stimulation (TENS) improves upper GI symptoms and balances the sympathovagal activity in scleroderma patients.* Dig Dis Sci, 2007. 52(5): p. 1329-37.

52.   Chandler, M.J., et al., *Effects of vagal afferent stimulation on cervical spinothalamic tract neurons in monkeys.* Pain, 1991. 44(1): p. 81-7.

53.   Chandler, M.J., J. Zhang, and R.D. Foreman, *Vagal, sympathetic and somatic sensory inputs to upper cervical (C1-C3) spinothalamic tract neurons in monkeys.* J Neurophysiol, 1996. 76(4): p. 2555-67.

54. Chandler, M.J., et al., *Spinal inhibitory effects of cardiopulmonary afferent inputs in monkeys: neuronal processing in high cervical segments.* J Neurophysiol, 2002. 87(3): p. 1290-302.

55. Weissman-Fogel, I., et al., *Vagal damage enhances polyneuropathy pain: additive effect of two algogenic mechanisms.* Pain, 2008. 138(1): p. 153-62.

56. Kwon, Y.H., et al., *Primary motor cortex activation by transcranial direct current stimulation in the human brain.* Neurosci Lett, 2008. 435(1): p. 56-9.

57. Nitsche, M.A., et al., *[Modulation of cortical excitability by transcranial direct current stimulation].* Nervenarzt, 2002. 73(4): p. 332-5.

58. Boros, K., et al., *Premotor transcranial direct current stimulation (tDCS) affects primary motor excitability in humans.* Eur J Neurosci, 2008. 27(5): p. 1292-300.

59. Fregni, F., et al., *Transcranial direct current stimulation.* Br J Psychiatry, 2005. 186: p. 446-7.

60. Jeffery, D.T., et al., *Effects of transcranial direct current stimulation on the excitability of the leg motor cortex.* Exp Brain Res, 2007. 182(2): p. 281-7.

61. Lang, N., et al., *How does transcranial DC stimulation of the primary motor cortex alter regional neuronal activity in the human brain?* Eur J Neurosci, 2005. 22(2): p. 495-504.

62. Alonso-Alonso, M., F. Fregni, and A. Pascual-Leone, *Brain stimulation in poststroke rehabilitation.* Cerebrovasc Dis, 2007. 24 Suppl 1: p. 157-66.

63. Boggio, P.S., et al., *Repeated sessions of noninvasive brain DC stimulation is associated with motor function improvement in stroke patients.* Restor Neurol Neurosci, 2007. 25(2): p. 123-9.

64. Brown, J.A., *Recovery of motor function after stroke.* Prog Brain Res, 2006. 157: p. 223-8.

65. Fregni, F., et al., *Transcranial direct current stimulation of the unaffected hemisphere in stroke patients.* Neuroreport, 2005. 16(14): p. 1551-5.

66. Boggio, P.S., et al., *Effects of transcranial direct current stimulation on working memory in patients with Parkinson's disease.* J Neurol Sci, 2006. 249(1): p. 31-8.

67. Fregni, F., et al., *Non-invasive brain stimulation for Parkinson's disease: a systematic review and meta-analysis of the literature.* J Neurol Neurosurg Psychiatry, 2005. 76(12): p. 1614-23.

68. Helmich, R.C., et al., *Repetitive transcranial magnetic stimulation to improve mood and motor function in Parkinson's disease.* J Neurol Sci, 2006. 248(1-2): p. 84-96.

69. Fregni, F., et al., *Cognitive effects of repeated sessions of transcranial direct current stimulation in patients with depression.* Depress Anxiety, 2006. 23(8): p. 482-4.

70. Nitsche, M.A., et al., *Facilitation of implicit motor learning by weak transcranial direct current stimulation of the primary motor cortex in the human.* J Cogn Neurosci, 2003. 15(4): p. 619-26.

71. Fregni, F., et al., *Anodal transcranial direct current stimulation of prefrontal cortex enhances working memory.* Exp Brain Res, 2005. 166(1): p. 23-30.

72. Lang, N., et al., *Effects of transcranial direct current stimulation over the human motor cortex on corticospinal and transcallosal excitability.* Exp Brain Res, 2004. 156(4): p. 439-43.

73. Roizenblatt, S., et al., *Site-specific effects of transcranial direct current stimulation on sleep and pain in fibromyalgia: a randomized, sham-controlled study.* Pain Pract, 2007. 7(4): p. 297-306.

74.   Thickbroom, G.W., *Transcranial magnetic stimulation and synaptic plasticity: experimental framework and human models.* Exp Brain Res, 2007. 180(4): p. 583-93.

75.   Tyc, F. and A. Boyadjian, *Cortical plasticity and motor activity studied with transcranial magnetic stimulation.* Rev Neurosci, 2006. 17(5): p. 469-95.

76.   Vandermeeren, Y., et al., *[Plasticity of motor maps in primates: recent advances and therapeutical perspectives].* Rev Neurol (Paris), 2003. 159(3): p. 259-75.

77.   Ward, N.S., *Neural plasticity and recovery of function.* Prog Brain Res, 2005. 150: p. 527-35.

78.   Ward, N.S., *Plasticity and the functional reorganization of the human brain.* Int J Psychophysiol, 2005. 58(2-3): p. 158-61.

79.   Ziemann, U., *TMS induced plasticity in human cortex.* Rev Neurosci, 2004. 15(4): p. 253-66.

80.   Poreisz, C., et al., *Safety aspects of transcranial direct current stimulation concerning healthy subjects and patients.* Brain Res Bull, 2007. 72(4-6): p. 208-14.

81.   Kanda, M., et al., *Primary somatosensory cortex is actively involved in pain processing in human.* Brain Res, 2000. 853(2): p. 282-9.

82.   Antal, A., et al., *Towards unravelling task-related modulations of neuroplastic changes induced in the human motor cortex.* Eur J Neurosci, 2007. 26(9): p. 2687-91.

83.   Nitsche, M.A., et al., *Shaping the effects of transcranial direct current stimulation of the human motor cortex.* J Neurophysiol, 2007. 97(4): p. 3109-17.

84.   Nitsche, M.A., et al., *Modulation of cortical excitability by weak direct current stimulation--technical, safety and functional aspects.* Suppl Clin Neurophysiol, 2003. 56: p. 255-76.

85.   Nitsche, M.A., et al., *Safety criteria for transcranial direct current stimulation (tDCS) in humans.* Clin Neurophysiol, 2003. 114(11): p. 2220-2; author reply 2222-3.

86.   Boggio, P.S., et al., *A randomized, double-blind clinical trial on the efficacy of cortical direct current stimulation for the treatment of major depression.* Int J Neuropsychopharmacol, 2008. 11(2): p. 249-54.

87.   Antal, A., M.A. Nitsche, and W. Paulus, *Transcranial magnetic and direct current stimulation of the visual cortex.* Suppl Clin Neurophysiol, 2003. 56: p. 291-304.

88.   Nitsche, M.A., et al., *Level of action of cathodal DC polarisation induced inhibition of the human motor cortex.* Clin Neurophysiol, 2003. 114(4): p. 600-4.

89.   Fregni, F., S. Freedman, and A. Pascual-Leone, *Recent advances in the treatment of chronic pain with non-invasive brain stimulation techniques.* Lancet Neurol, 2007. 6(2): p. 188-91.

90.   Fregni, F., et al., *A randomized, sham-controlled, proof of principle study of transcranial direct current stimulation for the treatment of pain in fibromyalgia.* Arthritis Rheum, 2006. 54(12): p. 3988-98.

91.   Kaptchuk, T., *The Web That Has No Weaver.* 2000, Toas: Redwing Book Company.

92.   Paulus, W., *Transcranial direct current stimulation (tDCS).* Suppl Clin Neurophysiol, 2003. 56: p. 249-54.

93.   Bae, C.S., et al., *Effect of Ga-as laser on the regeneration of injured sciatic nerves in the rat.* In Vivo, 2004. 18(4): p. 489-95.

94.   Elwakil, T.F., A. Elazzazi, and H. Shokeir, *Treatment of carpal tunnel syndrome by low-level laser versus open carpal tunnel release.* Lasers Med Sci, 2007. 22(4): p. 265-70.

95.   Miloro, M., et al., *Low-level laser effect on neural regeneration in Gore-Tex tubes.* Oral Surg Oral Med Oral Pathol Oral Radiol Endod, 2002. 93(1): p. 27-34.

96.   Mohammed, I.F., N. Al-Mustawfi, and L.N. Kaka, *Promotion of regenerative processes in injured peripheral nerve induced by low-level laser therapy.* Photomed Laser Surg, 2007. 25(2): p. 107-11.

97.   Peric, Z., *[Influence of low-intensity laser therapy on spatial perception threshold and electroneurographic finding in patients with diabetic polyneuropathy].* Srp Arh Celok Lek, 2007. 135(5-6): p. 257-63.

98.   Reddy, G.K., *Photobiological basis and clinical role of low-intensity lasers in biology and medicine.* J Clin Laser Med Surg, 2004. 22(2): p. 141-50.

99.   Rochkind, S., et al., *Efficacy of 780-nm laser phototherapy on peripheral nerve regeneration after neurotube reconstruction procedure (double-blind randomized study).* Photomed Laser Surg, 2007. 25(3): p. 137-43.

100.   Snyder, S.K., et al., *Quantitation of calcitonin gene-related peptide mRNA and neuronal cell death in facial motor nuclei following axotomy and 633 nm low power laser treatment.* Lasers Surg Med, 2002. 31(3): p. 216-22.

101.   Vinck, E., et al., *Evidence of changes in sural nerve conduction mediated by light emitting diode irradiation.* Lasers Med Sci, 2005. 20(1): p. 35-40.

102.   Whelan, H.T., et al., *Effect of NASA light-emitting diode irradiation on molecular changes for wound healing in diabetic mice.* J Clin Laser Med Surg, 2003. 21(2): p. 67-74.

103.   Zinman, L.H., et al., *Low-intensity laser therapy for painful symptoms of diabetic sensorimotor polyneuropathy: a controlled trial.* Diabetes Care, 2004. 27(4): p. 921-4.

104.   Miriutova, N.F., et al., *[Laser therapy and electric stimulation in rehabilitation treatment of peripheral neuropathy].* Vopr Kurortol Fizioter Lech Fiz Kult, 2002(4): p. 25-7.

105.   Saltmarche, A.E., *Low level laser therapy for healing acute and chronic wounds - the extendicare experience.* Int Wound J, 2008. 5(2): p. 351-60.

106.   Sobanko, J.F. and T.S. Alster, *Efficacy of Low-Level Laser Therapy for Chronic Cutaneous Ulceration in Humans: A Review and Discussion.* Dermatol Surg, 2008.

107.   Byrnes, K.R., et al., *Light promotes regeneration and functional recovery and alters the immune response after spinal cord injury.* Lasers Surg Med, 2005. 36(3): p. 171-85.

108.   Rochkind, S., et al., *Laser phototherapy (780 nm), a new modality in treatment of long-term incomplete peripheral nerve injury: a randomized double-blind placebo-controlled study.* Photomed Laser Surg, 2007. 25(5): p. 436-42.

109.   Currier, D.P., D. Greathouse, and T. Swift, *Sensory nerve conduction: effect of ultrasound.* Arch Phys Med Rehabil, 1978. 59(4): p. 181-5.

110.   Hsieh, Y.L., *Peripheral therapeutic ultrasound stimulation alters the distribution of spinal C-fos immunoreactivity induced by early or late phase of inflammation.* Ultrasound Med Biol, 2008. 34(3): p. 475-86.

111.   Hsieh, Y.L., *Reduction in induced pain by ultrasound may be caused by altered expression of spinal neuronal nitric oxide synthase-producing neurons.* Arch Phys Med Rehabil, 2005. 86(7): p. 1311-7.

112.   Schoffnegger, D., R. Ruscheweyh, and J. Sandkuhler, *Spread of excitation across modality borders in spinal dorsal horn of neuropathic rats.* Pain, 2008. 135(3): p. 300-10.

113.   Maihofner, C., H.O. Handwerker, and F. Birklein, *Functional imaging of allodynia in complex regional pain syndrome.* Neurology, 2006. 66(5): p. 711-7.

114.   Chang, C.J. and S.H. Hsu, *The effects of low-intensity ultrasound on peripheral nerve regeneration in poly(DL-lactic acid-co-glycolic acid) conduits seeded with Schwann cells.* Ultrasound Med Biol, 2004. 30(8): p. 1079-84.

115.   Crisci, A.R. and A.L. Ferreira, *Low-intensity pulsed ultrasound accelerates the regeneration of the sciatic nerve after neurotomy in rats.* Ultrasound Med Biol, 2002. 28(10): p. 1335-41.

116.   Lazar, D.A., et al., *Acceleration of recovery after injury to the peripheral nervous system using ultrasound and other therapeutic modalities.* Neurosurg Clin N Am, 2001. 12(2): p. 353-7.

117.   Paik, N.J., S.H. Cho, and T.R. Han, *Ultrasound therapy facilitates the recovery of acute pressure-induced conduction block of the median nerve in rabbits.* Muscle Nerve, 2002. 26(3): p. 356-61.

118.   Mourad, P.D., et al., *Ultrasound accelerates functional recovery after peripheral nerve damage.* Neurosurgery, 2001. 48(5): p. 1136-40; discussion 1140-1.

119.   Ahn, A.C., et al., *Two styles of acupuncture for treating painful diabetic neuropathy--a pilot randomised control trial.* Acupunct Med, 2007. 25(1-2): p. 11-7.

120.   Schroder, S., et al., *Acupuncture treatment improves nerve conduction in peripheral neuropathy.* Eur J Neurol, 2007. 14(3): p. 276-81.

121.   Lin, C.C., et al., *Chronic electrical stimulation of four acupuncture points on rat diabetic neuropathy.* Conf Proc IEEE Eng Med Biol Soc, 2005. 4: p. 4271-4.

122.   Dhond, R.P., N. Kettner, and V. Napadow, *Neuroimaging acupuncture effects in the human brain.* J Altern Complement Med, 2007. 13(6): p. 603-16.

123.   Dhond, R.P., et al., *Spatiotemporal mapping the neural correlates of acupuncture with MEG.* J Altern Complement Med, 2008. 14(6): p. 679-88.

124.   Napadow, V., et al., *Correlating acupuncture FMRI in the human brainstem with heart rate variability.* Conf Proc IEEE Eng Med Biol Soc, 2005. 5: p. 4496-9.

125.   Napadow, V., et al., *Hypothalamus and amygdala response to acupuncture stimuli in Carpal Tunnel Syndrome.* Pain, 2007. 130(3): p. 254-66.

126.   Napadow, V., et al., *Somatosensory cortical plasticity in carpal tunnel syndrome treated by acupuncture.* Hum Brain Mapp, 2007. 28(3): p. 159-71.

127.   Napadow, V., et al., *Effects of electroacupuncture versus manual acupuncture on the human brain as measured by fMRI.* Hum Brain Mapp, 2005. 24(3): p. 193-205.

128.   Inoue, M., et al., *Effects of lumbar acupuncture stimulation on blood flow to the sciatic nerve trunk--an exploratory study.* Acupunct Med, 2005. 23(4): p. 166-70.

129.   La, J.L., S. Jalali, and S.A. Shami, *Morphological studies on crushed sciatic nerve of rabbits with electroacupuncture or diclofenac sodium treatment.* Am J Chin Med, 2005. 33(4): p. 663-9.

130.	Phillips, K.D., W.D. Skelton, and G.A. Hand, *Effect of acupuncture administered in a group setting on pain and subjective peripheral neuropathy in persons with human immunodeficiency virus disease.* J Altern Complement Med, 2004. 10(3): p. 449-55.

131.	Wang, Y.P., et al., *[Effects of acupuncture on diabetic peripheral neuropathies].* Zhongguo Zhen Jiu, 2005. 25(8): p. 542-4.

132.	Wong, R. and S. Sagar, *Acupuncture treatment for chemotherapy-induced peripheral neuropathy--a case series.* Acupunct Med, 2006. 24(2): p. 87-91.

133.	Zhao, H.L., X. Gao, and Y.B. Gao, *[Clinical observation on effect of acupuncture in treating diabetic peripheral neuropathy].* Zhongguo Zhong Xi Yi Jie He Za Zhi, 2007. 27(4): p. 312-4.

134.	Zhao, J.L. and Z.R. Li, *[Clinical observation on mild-warm moxibustion for treatment of diabetic peripheral neuropathy].* Zhongguo Zhen Jiu, 2008. 28(1): p. 13-6.

135.	Krishnan, A.V., et al., *Sensory nerve excitability and neuropathy in end stage kidney disease.* J Neurol Neurosurg Psychiatry, 2006. 77(4): p. 548-51.

136.	Pitcher, G.M. and J.L. Henry, *Cellular mechanisms of hyperalgesia and spontaneous pain in a spinalized rat model of peripheral neuropathy: changes in myelinated afferent inputs implicated.* Eur J Neurosci, 2000. 12(6): p. 2006-20.

137.	Tamura, N., et al., *Increased nodal persistent Na+ currents in human neuropathy and motor neuron disease estimated by latent addition.* Clin Neurophysiol, 2006. 117(11): p. 2451-8.

138.	Sanada, M. and H. Yasuda, *[Role of Ca2+ channels in the pathogenesis of diabetic neuropathy].* Nippon Rinsho, 2005. 63 Suppl 6: p. 647-51.

139.	Sarantopoulos, C., et al., *Gabapentin decreases membrane calcium currents in injured as well as in control mammalian primary afferent neurons.* Reg Anesth Pain Med, 2002. 27(1): p. 47-57.

140.	Uenishi, H., et al., *Ion channel modulation as the basis for neuroprotective action of MS-153.* Ann N Y Acad Sci, 1999. 890: p. 385-99.

141.	Capiati, D.A., G. Vazquez, and R.L. Boland, *Protein kinase C alpha modulates the Ca2+ influx phase of the Ca2+ response to 1alpha,25-dihydroxy-vitamin-D3 in skeletal muscle cells.* Horm Metab Res, 2001. 33(4): p. 201-6.

142.	Curr Drug Targets, 2000. 1(2): p. 163-83.

143.	Casellini, C.M., et al., *A 6-month, randomized, double-masked, placebo-controlled study evaluating the effects of the protein kinase C-beta inhibitor ruboxistaurin on skin microvascular blood flow and other measures of diabetic peripheral neuropathy.* Diabetes Care, 2007. 30(4): p. 896-902.

144.	Mol Cell Biochem, 2007. 302(1-2): p. 179-85.

145.	Das Evcimen, N. and G.L. King, *The role of protein kinase C activation and the vascular complications of diabetes.* Pharmacol Res, 2007. 55(6): p. 498-510.

146.	Eichberg, J., *Protein kinase C changes in diabetes: is the concept relevant to neuropathy?* Int Rev Neurobiol, 2002. 50: p. 61-82.

147.	Joy, S.V., et al., *Ruboxistaurin, a protein kinase C beta inhibitor, as an emerging treatment for diabetes microvascular complications.* Ann Pharmacother, 2005. 39(10): p. 1693-9.

148.	Kamei, J., et al., *Therapeutic potential of PKC inhibitors in painful diabetic neuropathy.* Expert Opin Investig Drugs, 2001. 10(9): p. 1653-64.

149.   Kiss, L. and C. Szabo, *The pathogenesis of diabetic complications: the role of DNA injury and poly(ADP-ribose) polymerase activation in peroxynitrite-mediated cytotoxicity.* Mem Inst Oswaldo Cruz, 2005. 100 Suppl 1: p. 29-37.

150.   McCarty, M.F., *PKC-mediated modulation of L-type calcium channels may contribute to fat-induced insulin resistance.* Med Hypotheses, 2006. 66(4): p. 824-31.

151.   Shangguan, Y., et al., *Diabetic neuropathy: inhibitory G protein dysfunction involves PKC-dependent phosphorylation of Goalpha.* J Neurochem, 2003. 86(4): p. 1006-14.

152.   J Neurochem, 2008. 104(2): p. 491-9.

153.   Ahmed, N. and P.J. Thornalley, *Advanced glycation endproducts: what is their relevance to diabetic complications?* Diabetes Obes Metab, 2007. 9(3): p. 233-45.

154.   Beltramo, E., et al., *Thiamine and benfotiamine prevent increased apoptosis in endothelial cells and pericytes cultured in high glucose.* Diabetes Metab Res Rev, 2004. 20(4): p. 330-6.

155.   Koike, H., et al., *Alcoholic neuropathy is clinicopathologically distinct from thiamine-deficiency neuropathy.* Ann Neurol, 2003. 54(1): p. 19-29.

156.   Koike, H. and G. Sobue, *Alcoholic neuropathy.* Curr Opin Neurol, 2006. 19(5): p. 481-6.

157.   Thornalley, P.J., *The potential role of thiamine (vitamin B(1)) in diabetic complications.* Curr Diabetes Rev, 2005. 1(3): p. 287-98.

158.   Cunningham, J.J., P.L. Mearkle, and R.G. Brown, *Vitamin C: an aldose reductase inhibitor that normalizes erythrocyte sorbitol in insulin-dependent diabetes mellitus.* J Am Coll Nutr, 1994. 13(4): p. 344-50.

159.   Ortwerth, B.J., et al., *Ascorbic acid glycation: the reactions of L-threose in lens tissue.* Exp Eye Res, 1994. 58(6): p. 665-74.

160.   Perkins, B.A. and V. Bril, *Emerging therapies for diabetic neuropathy: a clinical overview.* Curr Diabetes Rev, 2005. 1(3): p. 271-80.

161.   Stevens, M.J., et al., *Osmotically-induced nerve taurine depletion and the compatible osmolyte hypothesis in experimental diabetic neuropathy in the rat.* Diabetologia, 1993. 36(7): p. 608-14.

162.   Hoybergs, Y.M. and T.F. Meert, *The effect of low-dose insulin on mechanical sensitivity and allodynia in type I diabetes neuropathy.* Neurosci Lett, 2007. 417(2): p. 149-54.

163.   Kamiya, H., et al., *C-Peptide reverses nociceptive neuropathy in type 1 diabetes.* Diabetes, 2006. 55(12): p. 3581-7.

164.   Toth, C., et al., *Diabetes mellitus and the sensory neuron.* J Neuropathol Exp Neurol, 2004. 63(6): p. 561-73.

165.   Toth, C., et al., *Rescue and regeneration of injured peripheral nerve axons by intrathecal insulin.* Neuroscience, 2006. 139(2): p. 429-49.

166.   Zhang, W., et al., *Human C-peptide dose dependently prevents early neuropathy in the BB/Wor-rat.* Int J Exp Diabetes Res, 2001. 2(3): p. 187-93.

167.   Apfel, S.C. and J.A. Kessler, *Neurotrophic factors in the therapy of peripheral neuropathy.* Baillieres Clin Neurol, 1995. 4(3): p. 593-606.

168. Apfel, S.C. and J.A. Kessler, *Neurotrophic factors in the treatment of peripheral neuropathy.* Ciba Found Symp, 1996. 196: p. 98-108; discussion 108-12.

169. Azar, Z.M., M.Z. Mehdi, and A.K. Srivastava, Can J Physiol Pharmacol, 2007. 85(1): p. 105-11.

170. Ditchkoff, S.S., et al., *Concentrations of insulin-like growth factor-I in adult male white-tailed deer (Odocoileus virginianus): associations with serum testosterone, morphometrics and age during and after the breeding season.* Comp Biochem Physiol A Mol Integr Physiol, 2001. 129(4): p. 887-95.

171. Ishii, D.N., *Implication of insulin-like growth factors in the pathogenesis of diabetic neuropathy.* Brain Res Brain Res Rev, 1995. 20(1): p. 47-67.

172. Ishii, D.N. and S.B. Lupien, *Insulin-like growth factors protect against diabetic neuropathy: effects on sensory nerve regeneration in rats.* J Neurosci Res, 1995. 40(1): p. 138-44.

173. Lewis, M.E., et al., *Insulin-like growth factor-I: potential for treatment of motor neuronal disorders.* Exp Neurol, 1993. 124(1): p. 73-88.

174. Migdalis, I.N., et al., *Insulin-like growth factor-I and IGF-I receptors in diabetic patients with neuropathy.* Diabet Med, 1995. 12(9): p. 823-7.

175. Horm Metab Res, 2008. 40(3): p. 163-4.

176. Sadighi, M., et al., *Effects of insulin-like growth factor-I (IGF-I) and IGF-II on the growth of antler cells in vitro.* J Endocrinol, 1994. 143(3): p. 461-9.

177. Shavlakadze, T., et al., *Insulin-like growth factor I slows the rate of denervation induced skeletal muscle atrophy.* Neuromuscul Disord, 2005. 15(2): p. 139-46.

178. Vaught, J.L., et al., *Potential utility of rhIGF-1 in neuromuscular and/or degenerative disease.* Ciba Found Symp, 1996. 196: p. 18-27; discussion 27-38.

179. Sima, A.A., et al., *Molecular alterations underlie nodal and paranodal degeneration in type 1 diabetic neuropathy and are prevented by C-peptide.* Diabetes, 2004. 53(6): p. 1556-63.

180. Hwang, S.W. and U. Oh, *Current concepts of nociception: nociceptive molecular sensors in sensory neurons.* Curr Opin Anaesthesiol, 2007. 20(5): p. 427-34.

181. Kempuraj, D., et al., *Flavonols inhibit proinflammatory mediator release, intracellular calcium ion levels and protein kinase C theta phosphorylation in human mast cells.* Br J Pharmacol, 2005. 145(7): p. 934-44.

182. *Ruboxistaurin: LY 333531.* Drugs R D, 2007. 8(3): p. 193-9.

183. Ahlgren, S.C. and J.D. Levine, *Protein kinase C inhibitors decrease hyperalgesia and C-fiber hyperexcitability in the streptozotocin-diabetic rat.* J Neurophysiol, 1994. 72(2): p. 684-92.

184. Bursell, S.E. and G.L. King, *Can protein kinase C inhibition and vitamin E prevent the development of diabetic vascular complications?* Diabetes Res Clin Pract, 1999. 45(2-3): p. 169-82.

185. Cameron, N.E. and M.A. Cotter, *Effects of protein kinase Cbeta inhibition on neurovascular dysfunction in diabetic rats: interaction with oxidative stress and essential fatty acid dysmetabolism.* Diabetes Metab Res Rev, 2002. 18(4): p. 315-23.

186. Bolcskei, K., et al., *Investigation of the role of TRPV1 receptors in acute and chronic nociceptive processes using gene-deficient mice.* Pain, 2005. 117(3): p. 368-76.

187.   ]Bastianetto, S., W.H. Zheng, and R. Quirion, *The Ginkgo biloba extract (EGb 761) protects and rescues hippocampal cells against nitric oxide-induced toxicity: involvement of its flavonoid constituents and protein kinase C.* J Neurochem, 2000. 74(6): p. 2268-77.

188.   Bastianetto, S. and R. Quirion, *Natural extracts as possible protective agents of brain aging.* Neurobiol Aging, 2002. 23(5): p. 891-97.

189.   Azzi, A., *The role of alpha-tocopherol in preventing disease.* Eur J Nutr, 2004. 43 Suppl 1: p. I/18-25.

190.   Carter, C.A. and C.J. Kane, *Therapeutic potential of natural compounds that regulate the activity of protein kinase C.* Curr Med Chem, 2004. 11(21): p. 2883-902.

191.   Sylvester, P.W., et al., *Vitamin E inhibition of normal mammary epithelial cell growth is associated with a reduction in protein kinase C(alpha) activation.* Cell Prolif, 2001. 34(6): p. 347-57.

192.   Karihaloo, A.K., K. Joshi, and J.S. Chopra, *Effect of sorbinil and ascorbic acid on myo-inositol transport in cultured rat Schwann cells exposed to elevated extracellular glucose.* J Neurochem, 1997. 69(5): p. 2011-8.

193.   Cameron, N.E. and M.A. Cotter, *Comparison of the effects of ascorbyl gamma-linolenic acid and gamma-linolenic acid in the correction of neurovascular deficits in diabetic rats.* Diabetologia, 1996. 39(9): p. 1047-54.

194.   Wang, H., et al., *Experimental and clinical studies on the reduction of erythrocyte sorbitol-glucose ratios by ascorbic acid in diabetes mellitus.* Diabetes Res Clin Pract, 1995. 28(1): p. 1-8.

195.   Jung, H.A., et al., *Inhibitory activities of prenylated flavonoids from Sophora flavescens against aldose reductase and generation of advanced glycation endproducts.* J Pharm Pharmacol, 2008. 60(9): p. 1227-36.

196.   Kato, A., et al., *Protective Effects of Dietary Chamomile Tea on Diabetic Complications.* J Agric Food Chem, 2008.

197.   Kim, J.M., et al., *Aldose-reductase- and protein-glycation-inhibitory principles from the whole plant of Duchesnea chrysantha.* Chem Biodivers, 2008. 5(2): p. 352-6.

198.   Ramana, B.V., et al., *Effect of quercetin on galactose-induced hyperglycaemic oxidative stress in hepatic and neuronal tissues of Wistar rats.* Acta Diabetol, 2006. 43(4): p. 135-41.

199.   Ramana, B.V., et al., *Defensive role of quercetin against imbalances of calcium, sodium, and potassium in galactosemic cataract.* Biol Trace Elem Res, 2007. 119(1): p. 35-41.

200.   Varma, S.D., I. Mikuni, and J.H. Kinoshita, *Flavonoids as inhibitors of lens aldose reductase.* Science, 1975. 188(4194): p. 1215-6.

201.   Ang, C.D., et al., *Vitamin B for treating peripheral neuropathy.* Cochrane Database Syst Rev, 2008(3): p. CD004573.

202.   Giusti, C. and P. Gargiulo, *Advances in biochemical mechanisms of diabetic retinopathy.* Eur Rev Med Pharmacol Sci, 2007. 11(3): p. 155-63.

203.   Jermendy, G., *Evaluating thiamine deficiency in patients with diabetes.* Diab Vasc Dis Res, 2006. 3(2): p. 120-1.

204.   Medvedeva, L.A., et al., *[Neurometabolic therapy of diabetic neuropathy].* Zh Nevrol Psikhiatr Im S S Korsakova, 2006. 106(7): p. 71-3.

205. Varkonyi, T. and P. Kempler, *Diabetic neuropathy: new strategies for treatment.* Diabetes Obes Metab, 2008. 10(2): p. 99-108.

206. Lonsdale, D., *Thiamine tetrahydrofurfuryl disulfide: a little known therapeutic agent.* Med Sci Monit, 2004. 10(9): p. RA199-203.

207. Baker, H. and O. Frank, *Absorption, utilization and clinical effectiveness of allithiamines compared to water-soluble thiamines.* J Nutr Sci Vitaminol (Tokyo), 1976. 22 SUPPL: p. 63-8.

208. Beltramo, E., et al., *Effects of thiamine and benfotiamine on intracellular glucose metabolism and relevance in the prevention of diabetic complications.* Acta Diabetol, 2008. 45(3): p. 131-41.

209. Du, X., D. Edelstein, and M. Brownlee, *Oral benfotiamine plus alpha-lipoic acid normalises complication-causing pathways in type 1 diabetes.* Diabetologia, 2008.

210. Marchetti, V., et al., *Benfotiamine counteracts glucose toxicity effects on endothelial progenitor cell differentiation via Akt/FoxO signaling.* Diabetes, 2006. 55(8): p. 2231-7.

211. Sanchez-Ramirez, G.M., et al., *Benfotiamine relieves inflammatory and neuropathic pain in rats.* Eur J Pharmacol, 2006. 530(1-2): p. 48-53.

212. Stirban, A., et al., *Benfotiamine prevents macro- and microvascular endothelial dysfunction and oxidative stress following a meal rich in advanced glycation end products in individuals with type 2 diabetes.* Diabetes Care, 2006. 29(9): p. 2064-71.

213. Stracke, H., et al., *Benfotiamine in Diabetic Polyneuropathy (BENDIP): Results of a Randomised, Double Blind, Placebo-controlled Clinical Study.* Exp Clin Endocrinol Diabetes, 2008.

214. Wu, S. and J. Ren, *Benfotiamine alleviates diabetes-induced cerebral oxidative damage independent of advanced glycation end-product, tissue factor and TNF-alpha.* Neurosci Lett, 2006. 394(2): p. 158-62.

215. Mohamed-Ali, V. and J. Pinkney, *Therapeutic potential of insulin-like growth factor-1 in patients with diabetes mellitus.* Treat Endocrinol, 2002. 1(6): p. 399-410.

216. Dore, S., et al., *Protective and rescuing abilities of IGF-I and some putative free radical scavengers against beta-amyloid-inducing toxicity in neurons.* Ann N Y Acad Sci, 1999. 890: p. 356-64.

217. Gu, L., et al., *Expression and localization of insulin-like growth factor-I in four parts of the red deer antler.* Growth Factors, 2007. 25(4): p. 264-79.

218. Elliott, J.L., et al., *Effect of testosterone on binding of insulin-like growth factor-I (IGF-I) and IGF-II in growing antlers of fallow deer (Dama dama).* Growth Regul, 1996. 6(4): p. 214-21.

219. Suttie, J.M., et al., *Pulsatile growth hormone, insulin-like growth factors and antler development in red deer (Cervus elaphus scoticus) stags.* J Endocrinol, 1989. 121(2): p. 351-60.

220. Soumyanath, A., et al., *Centella asiatica accelerates nerve regeneration upon oral administration and contains multiple active fractions increasing neurite elongation in-vitro.* J Pharm Pharmacol, 2005. 57(9): p. 1221-9.

221. Fantus, I.G., et al., *Modulation of insulin action by vanadate: evidence of a role for phosphotyrosine phosphatase activity to alter cellular signaling.* Mol Cell Biochem, 1995. 153(1-2): p. 103-12.

222. Kadota, S., et al., *Vanadate stimulation of IGF binding to rat adipocytes.* Biochem Biophys Res Commun, 1986. 138(1): p. 174-8.

223.   Mehdi, M.Z., et al., *Involvement of insulin-like growth factor type 1 receptor and protein kinase Cdelta in bis(maltolato)oxovanadium(IV)-induced phosphorylation of protein kinase B in HepG2 cells.* Biochemistry, 2006. 45(38): p. 11605-15.

224.   Basuki, W., et al., *Enhancement of insulin signaling pathway in adipocytes by oxovanadium(IV) complexes.* Biochem Biophys Res Commun, 2006. 349(3): p. 1163-70.

225.   Nakai, M., et al., *Synthesis and insulinomimetic activities of novel mono- and tetranuclear oxovanadium(IV) complexes with 3-hydroxypyridine-2-carboxylic acid.* J Inorg Biochem, 2004. 98(1): p. 105-12.

226.   Srivastava, A.K. and M.Z. Mehdi, *Insulino-mimetic and anti-diabetic effects of vanadium compounds.* Diabet Med, 2005. 22(1): p. 2-13.

227.   Thompson, K.H. and C. Orvig, *Vanadium in diabetes: 100 years from Phase 0 to Phase I.* J Inorg Biochem, 2006. 100(12): p. 1925-35.

228.   Jin, H.Y., et al., *The effect of alpha-lipoic acid on symptoms and skin blood flow in diabetic neuropathy.* Diabet Med, 2007. 24(9): p. 1034-8.

229.   Becic, F., E. Kapic, and M. Rakanovic-Todic, *[Pharmacological significance of alpha lipoic acid in up to date treatment of diabetic neuropathy].* Med Arh, 2008. 62(1): p. 45-8.

230.   Konrad, D., *Utilization of the insulin-signaling network in the metabolic actions of alpha-lipoic acid-reduction or oxidation?* Antioxid Redox Signal, 2005. 7(7-8): p. 1032-9.

231.   Liu, F., et al., *[Curative effect of alpha-lipoic acid on peripheral neuropathy in type 2 diabetes: a clinical study].* Zhonghua Yi Xue Za Zhi, 2007. 87(38): p. 2706-9.

232.   Ruessmann, H.J., *Switching from pathogenetic treatment with alpha-lipoic acid to gabapentin and other analgesics in painful diabetic neuropathy: a real-world study in outpatients.* J Diabetes Complications, 2008.

233.   Al'-Zamil, M.K. and E.V. Brezhneva, *[Implication of alpha-lipoic acid preparations in the treatment of diabetic neuropathy.].* Zh Nevrol Psikhiatr Im S S Korsakova, 2008. 108(2): p. 27-30.

234.   Ziegler, D., *Thioctic acid for patients with symptomatic diabetic polyneuropathy: a critical review.* Treat Endocrinol, 2004. 3(3): p. 173-89.

235.   Foster, T.S., *Efficacy and safety of alpha-lipoic acid supplementation in the treatment of symptomatic diabetic neuropathy.* Diabetes Educ, 2007. 33(1): p. 111-7.

236.   Choi, S.J., et al., *Effects of a polyacetylene from Panax ginseng on Na+ currents in rat dorsal root ganglion neurons.* Brain Res, 2008. 1191: p. 75-83.

237.   Shin, Y.H., et al., *Ginsenosides that produce differential antinociception in mice.* Gen Pharmacol, 1999. 32(6): p. 653-9.

238.   Tachikawa, E., et al., *Ginseng saponins reduce acetylcholine-evoked Na+ influx and catecholamine secretion in bovine adrenal chromaffin cells.* J Pharmacol Exp Ther, 1995. 273(2): p. 629-36.

239.   Choi, S., et al., *Effect of ginsenosides on voltage-dependent Ca(2+) channel subtypes in bovine chromaffin cells.* J Ethnopharmacol, 2001. 74(1): p. 75-81.

240.   Kim, H.S., et al., *Effects of ginsenosides on Ca2+ channels and membrane capacitance in rat adrenal chromaffin cells.* Brain Res Bull, 1998. 46(3): p. 245-51.

241.   Kim, S., S.Y. Nah, and H. Rhim, *Neuroprotective effects of ginseng saponins against L-type Ca2+ channel-mediated cell death in rat cortical neurons.* Biochem Biophys Res Commun, 2008. 365(3): p. 399-405.

242.   Nah, S.Y. and E.W. McCleskey, *Ginseng root extract inhibits calcium channels in rat sensory neurons through a similar path, but different receptor, as mu-type opioids.* J Ethnopharmacol, 1994. 42(1): p. 45-51.

243.   Nah, S.Y., H.J. Park, and E.W. McCleskey, *A trace component of ginseng that inhibits Ca2+ channels through a pertussis toxin-sensitive G protein.* Proc Natl Acad Sci U S A, 1995. 92(19): p. 8739-43.

244.   Liu, D., et al., *Voltage-dependent inhibition of brain Na(+) channels by American ginseng.* Eur J Pharmacol, 2001. 413(1): p. 47-54.

245.   Choi, S.E., et al., *Effects of ginsenosides on GABA(A) receptor channels expressed in Xenopus oocytes.* Arch Pharm Res, 2003. 26(1): p. 28-33.

246.   Chiechio, S., et al., *Acetyl-L-carnitine in neuropathic pain: experimental data.* CNS Drugs, 2007. 21 Suppl 1: p. 31-8; discussion 45-6.

247.   De Grandis, D., *Acetyl-L-carnitine for the treatment of chemotherapy-induced peripheral neuropathy: a short review.* CNS Drugs, 2007. 21 Suppl 1: p. 39-43; discussion 45-6.

248.   Di Cesare Mannelli, L., et al., *Protective effect of acetyl-L-carnitine on the apoptotic pathway of peripheral neuropathy.* Eur J Neurosci, 2007. 26(4): p. 820-7.

249.   Gonzalez-Duarte, A., K. Cikurel, and D.M. Simpson, *Managing HIV peripheral neuropathy.* Curr HIV/AIDS Rep, 2007. 4(3): p. 114-8.

250.   Xiao, W.H. and G.J. Bennett, *Chemotherapy-evoked neuropathic pain: Abnormal spontaneous discharge in A-fiber and C-fiber primary afferent neurons and its suppression by acetyl-L-carnitine.* Pain, 2008. 135(3): p. 262-70.

251.   Stillman, M. and J.P. Cata, *Management of chemotherapy-induced peripheral neuropathy.* Curr Pain Headache Rep, 2006. 10(4): p. 279-87.

252.   Cavaliere, R. and D. Schiff, *Neurologic toxicities of cancer therapies.* Curr Neurol Neurosci Rep, 2006. 6(3): p. 218-26.

253.   Bianchi, G., et al., *Symptomatic and neurophysiological responses of paclitaxel- or cisplatin-induced neuropathy to oral acetyl-L-carnitine.* Eur J Cancer, 2005. 41(12): p. 1746-50.

254.   Flatters, S.J., W.H. Xiao, and G.J. Bennett, *Acetyl-L-carnitine prevents and reduces paclitaxel-induced painful peripheral neuropathy.* Neurosci Lett, 2006. 397(3): p. 219-23.

255.   Ghirardi, O., et al., *Acetyl-L-Carnitine prevents and reverts experimental chronic neurotoxicity induced by oxaliplatin, without altering its antitumor properties.* Anticancer Res, 2005. 25(4): p. 2681-7.

256.   Ghirardi, O., et al., *Chemotherapy-induced allodinia: neuroprotective effect of acetyl-L-carnitine.* In Vivo, 2005. 19(3): p. 631-7.

257.   Herzmann, C., M.A. Johnson, and M. Youle, *Long-term effect of acetyl-L-carnitine for antiretroviral toxic neuropathy.* HIV Clin Trials, 2005. 6(6): p. 344-50.

258.   Jin, H.W., et al., *Prevention of paclitaxel-evoked painful peripheral neuropathy by acetyl-L-carnitine: effects on axonal mitochondria, sensory nerve fiber terminal arbors, and cutaneous Langerhans cells.* Exp Neurol, 2008. 210(1): p. 229-37.

259.   Joseph, E.K., et al., *Oxaliplatin acts on IB4-positive nociceptors to induce an oxidative stress-dependent acute painful peripheral neuropathy.* J Pain, 2008. 9(5): p. 463-72.

260.   Maestri, A., et al., *A pilot study on the effect of acetyl-L-carnitine in paclitaxel- and cisplatin-induced peripheral neuropathy.* Tumori, 2005. 91(2): p. 135-8.

261.   Osio, M., et al., *Acetyl-l-carnitine in the treatment of painful antiretroviral toxic neuropathy in human immunodeficiency virus patients: an open label study.* J Peripher Nerv Syst, 2006. 11(1): p. 72-6.

262.   Youle, M., *Acetyl-L-carnitine in HIV-associated antiretroviral toxic neuropathy.* CNS Drugs, 2007. 21 Suppl 1: p. 25-30; discussion 45-6.

263.   Youle, M. and M. Osio, *A double-blind, parallel-group, placebo-controlled, multicentre study of acetyl L-carnitine in the symptomatic treatment of antiretroviral toxic neuropathy in patients with HIV-1 infection.* HIV Med, 2007. 8(4): p. 241-50.

264.   Ohsawa, M., et al., *Preventive effect of acetyl-L-carnitine on the thermal hypoalgesia in streptozotocin-induced diabetic mice.* Eur J Pharmacol, 2008. 588(2-3): p. 213-6.

265.   Poorabbas, A., et al., *Determination of free L-carnitine levels in type II diabetic women with and without complications.* Eur J Clin Nutr, 2007. 61(7): p. 892-5.

266.   Sima, A.A., *The heterogeneity of diabetic neuropathy.* Front Biosci, 2008. 13: p. 4809-16.

267.   Uzun, N., et al., *Peripheric and automatic neuropathy in children with type 1 diabetes mellitus: the effect of L-carnitine treatment on the peripheral and autonomic nervous system.* Electromyogr Clin Neurophysiol, 2005. 45(6): p. 343-51.

268.   Kotil, K., et al., *Neuroprotective effects of acetyl-L-carnithine in experimental chronic compression neuropathy. A prospective, randomized and placebo-control trials.* Turk Neurosurg, 2007. 17(2): p. 67-77.

269.   Rowley, T.J., et al., *The antinociceptive response to nicotinic agonists in a mouse model of postoperative pain.* Anesth Analg, 2008. 107(3): p. 1052-7.

270.   Ivy Carroll, F., et al., *Synthesis, nicotinic acetylcholine receptor binding, antinociceptive and seizure properties of methyllycaconitine analogs.* Bioorg Med Chem, 2007. 15(2): p. 678-85.

271.   Cucchiaro, G., N. Chaijale, and K.G. Commons, *The locus coeruleus nucleus as a site of action of the antinociceptive and behavioral effects of the nicotinic receptor agonist, epibatidine.* Neuropharmacology, 2006. 50(7): p. 769-76.

272.   Carroll, F.I., et al., *Synthesis, nicotinic acetylcholine receptor binding, and antinociceptive properties of 3'-substituted deschloroepibatidine analogues. Novel nicotinic antagonists.* J Med Chem, 2005. 48(4): p. 1221-8.

273.   Damaj, M.I., et al., *Antinociceptive and pharmacological effects of metanicotine, a selective nicotinic agonist.* J Pharmacol Exp Ther, 1999. 291(1): p. 390-8.

274.   Decker, M.W., et al., *Antinociceptive effects of the novel neuronal nicotinic acetylcholine receptor agonist, ABT-594, in mice.* Eur J Pharmacol, 1998. 346(1): p. 23-33.

275.   Di Cesare Mannelli, L., et al., *Neuroprotective effects of acetyl-L-carnitine on neuropathic pain and apoptosis: A role for the nicotinic receptor.* J Neurosci Res, 2008.

276.   Kennedy, D.O. and A.B. Scholey, *The psychopharmacology of European herbs with cognition-enhancing properties.* Curr Pharm Des, 2006. 12(35): p. 4613-23.

277.   Kennedy, D.O., et al., *Modulation of mood and cognitive performance following acute administration of Melissa officinalis (lemon balm).* Pharmacol Biochem Behav, 2002. 72(4): p. 953-64.

278.   Kennedy, D.O., et al., *Modulation of mood and cognitive performance following acute administration of single doses of Melissa officinalis (Lemon balm) with human CNS nicotinic and muscarinic receptor-binding properties.* Neuropsychopharmacology, 2003. 28(10): p. 1871-81.

279.   Chen, W.Q., et al., *Role of taurine in regulation of intracellular calcium level and neuroprotective function in cultured neurons.* J Neurosci Res, 2001. 66(4): p. 612-9.

280.   Foos, T.M. and J.Y. Wu, *The role of taurine in the central nervous system and the modulation of intracellular calcium homeostasis.* Neurochem Res, 2002. 27(1-2): p. 21-6.

281.   Wu, J., et al., *Taurine activates glycine and gamma-aminobutyric acid A receptors in rat substantia gelatinosa neurons.* Neuroreport, 2008. 19(3): p. 333-7.

282.   Wu, Z.Y. and T.L. Xu, *Taurine-evoked chloride current and its potentiation by intracellular Ca2+ in immature rat hippocampal CA1 neurons.* Amino Acids, 2003. 24(1-2): p. 155-61.

283.   Pop-Busui, R., et al., *Depletion of taurine in experimental diabetic neuropathy: implications for nerve metabolic, vascular, and functional deficits.* Exp Neurol, 2001. 168(2): p. 259-72.

284.   Obrosova, I.G., L. Fathallah, and M.J. Stevens, *Taurine counteracts oxidative stress and nerve growth factor deficit in early experimental diabetic neuropathy.* Exp Neurol, 2001. 172(1): p. 211-9.

285.   Xiao, C., A. Giacca, and G.F. Lewis, *Oral taurine but not N-acetylcysteine ameliorates NEFA-induced impairment in insulin sensitivity and beta cell function in obese and overweight, non-diabetic men.* Diabetologia, 2008. 51(1): p. 139-46.

286.   Knabl, J., et al., *Reversal of pathological pain through specific spinal GABAA receptor subtypes.* Nature, 2008. 451(7176): p. 330-4.

287.   Naik, A.K., S. Pathirathna, and V. Jevtovic-Todorovic, *GABA(A) receptor modulation in dorsal root ganglia in vivo affects chronic pain after nerve injury.* Neuroscience, 2008. 154(4): p. 1539-53.

288.   Awad, R., et al., *Effects of traditionally used anxiolytic botanicals on enzymes of the gamma-aminobutyric acid (GABA) system.* Can J Physiol Pharmacol, 2007. 85(9): p. 933-42.

289.   Salah, S.M. and A.K. Jager, *Screening of traditionally used Lebanese herbs for neurological activities.* J Ethnopharmacol, 2005. 97(1): p. 145-9.

290.   Abuhamdah, S., et al., *Pharmacological profile of an essential oil derived from Melissa officinalis with anti-agitation properties: focus on ligand-gated channels.* J Pharm Pharmacol, 2008. 60(3): p. 377-84.

291.   Brown, E., et al., *Evaluation of the anxiolytic effects of chrysin, a Passiflora incarnata extract, in the laboratory rat.* AANA J, 2007. 75(5): p. 333-7.

292.   Zanoli, P., R. Avallone, and M. Baraldi, *Behavioral characterisation of the flavonoids apigenin and chrysin.* Fitoterapia, 2000. 71 Suppl 1: p. S117-23.

293.   Gramowski, A., et al., *Functional screening of traditional antidepressants with primary cortical neuronal networks grown on multielectrode neurochips.* Eur J Neurosci, 2006. 24(2): p. 455-65.

294.   Ichimura, T., et al., *Antihypertensive effect of an extract of Passiflora edulis rind in spontaneously hypertensive rats.* Biosci Biotechnol Biochem, 2006. 70(3): p. 718-21.

295.   Lolli, L.F., et al., *Possible involvement of GABA A-benzodiazepine receptor in the anxiolytic-like effect induced by Passiflora actinia extracts in mice.* J Ethnopharmacol, 2007. 111(2): p. 308-14.

296.   Nassiri-Asl, M., S. Shariati-Rad, and F. Zamansoltani, *Anticonvulsant effects of aerial parts of Passiflora incarnata extract in mice: involvement of benzodiazepine and opioid receptors.* BMC Complement Altern Med, 2007. 7: p. 26.

297.   Komori, T., et al., *The sleep-enhancing effect of valerian inhalation and sleep-shortening effect of lemon inhalation.* Chem Senses, 2006. 31(8): p. 731-7.

298.   Houghton, P.J., *The scientific basis for the reputed activity of Valerian.* J Pharm Pharmacol, 1999. 51(5): p. 505-12.

299.   Benke, D., et al., *GABA(A) receptors as in vivo substrate for the anxiolytic action of valerenic acid, a major constituent of valerian root extracts.* Neuropharmacology, 2008.

300.   Khom, S., et al., *Valerenic acid potentiates and inhibits GABA(A) receptors: molecular mechanism and subunit specificity.* Neuropharmacology, 2007. 53(1): p. 178-87.

301.   Ortiz, J.G., J. Nieves-Natal, and P. Chavez, *Effects of Valeriana officinalis extracts on [3H]flunitrazepam binding, synaptosomal [3H]GABA uptake, and hippocampal [3H]GABA release.* Neurochem Res, 1999. 24(11): p. 1373-8.

302.   Yuan, C.S., et al., *The gamma-aminobutyric acidergic effects of valerian and valerenic acid on rat brainstem neuronal activity.* Anesth Analg, 2004. 98(2): p. 353-8, table of contents.

303.   Incandela, L., et al., *Treatment of diabetic microangiopathy and edema with total triterpenic fraction of Centella asiatica: a prospective, placebo-controlled randomized study.* Angiology, 2001. 52 Suppl 2: p. S27-31.

304.   de Almeida, O.M., et al., *Opposite roles of GABA and excitatory amino acids on the control of GAD expression in cultured retina cells.* Brain Res, 2002. 925(1): p. 89-99.

305.   Erecinska, M., et al., *Regulation of GABA level in rat brain synaptosomes: fluxes through enzymes of the GABA shunt and effects of glutamate, calcium, and ketone bodies.* J Neurochem, 1996. 67(6): p. 2325-34.

306.   Fenalti, G., et al., *GABA production by glutamic acid decarboxylase is regulated by a dynamic catalytic loop.* Nat Struct Mol Biol, 2007. 14(4): p. 280-6.

307.   Fernandez-Pascual, S., et al., *Conversion into GABA (gamma-aminobutyric acid) may reduce the capacity of L-glutamine as an insulin secretagogue.* Biochem J, 2004. 379(Pt 3): p. 721-9.

308.   Pinal, C.S. and A.J. Tobin, *Uniqueness and redundancy in GABA production.* Perspect Dev Neurobiol, 1998. 5(2-3): p. 109-18.

309.   Amara, S., *Oral glutamine for the prevention of chemotherapy-induced peripheral neuropathy.* Ann Pharmacother, 2008. 42(10): p. 1481-5.

310.   Adam, F., et al., *A single preoperative dose of gabapentin (800 milligrams) does not augment postoperative analgesia in patients given interscalene brachial plexus blocks for arthroscopic shoulder surgery.* Anesth Analg, 2006. 103(5): p. 1278-82.

311.   Makino, H., *Treatment and care of neurotoxicity from taxane anticancer agents.* Breast Cancer, 2004. 11(1): p. 100-4.

312. Savarese, D.M., et al., *Prevention of chemotherapy and radiation toxicity with glutamine.* Cancer Treat Rev, 2003. 29(6): p. 501-13.

313. Strasser, F., et al., *Prevention of docetaxel- or paclitaxel-associated taste alterations in cancer patients with oral glutamine: a randomized, placebo-controlled, double-blind study.* Oncologist, 2008. 13(3): p. 337-46.

314. Stubblefield, M.D., et al., *Glutamine as a neuroprotective agent in high-dose paclitaxel-induced peripheral neuropathy: a clinical and electrophysiologic study.* Clin Oncol (R Coll Radiol), 2005. 17(4): p. 271-6.

315. Vahdat, L., et al., *Reduction of paclitaxel-induced peripheral neuropathy with glutamine.* Clin Cancer Res, 2001. 7(5): p. 1192-7.

316. Visovsky, C., et al., *Putting evidence into practice: evidence-based interventions for chemotherapy-induced peripheral neuropathy.* Clin J Oncol Nurs, 2007. 11(6): p. 901-13.

317. Wang, W.S., et al., *Oral glutamine is effective for preventing oxaliplatin-induced neuropathy in colorectal cancer patients.* Oncologist, 2007. 12(3): p. 312-9.

318. Wolf, S., et al., *Chemotherapy-induced peripheral neuropathy: prevention and treatment strategies.* Eur J Cancer, 2008. 44(11): p. 1507-15.

319. Arwert, L.I., J.B. Deijen, and M.L. Drent, *Effects of an oral mixture containing glycine, glutamine and niacin on memory, GH and IGF-I secretion in middle-aged and elderly subjects.* Nutr Neurosci, 2003. 6(5): p. 269-75.

320. Shao, A. and J.N. Hathcock, *Risk assessment for the amino acids taurine, L-glutamine and L-arginine.* Regul Toxicol Pharmacol, 2008. 50(3): p. 376-99.

321. Ward, E., et al., *Oral glutamine in paediatric oncology patients: a dose finding study.* Eur J Clin Nutr, 2003. 57(1): p. 31-6.

322. Ziegler, T.R., et al., *Safety and metabolic effects of L-glutamine administration in humans.* JPEN J Parenter Enteral Nutr, 1990. 14(4 Suppl): p. 137S-146S.

323. Zhou, X., et al., *Vinpocetine is a potent blocker of rat NaV1.8 tetrodotoxin-resistant sodium channels.* J Pharmacol Exp Ther, 2003. 306(2): p. 498-504.

324. Nekrassov, V. and M. Sitges, *Vinpocetine inhibits the epileptic cortical activity and auditory alterations induced by pentylenetetrazole in the guinea pig in vivo.* Epilepsy Res, 2004. 60(1): p. 63-71.

325. Sitges, M., E. Galvan, and V. Nekrassov, *Vinpocetine blockade of sodium channels inhibits the rise in sodium and calcium induced by 4-aminopyridine in synaptosomes.* Neurochem Int, 2005. 46(7): p. 533-40.

326. Sitges, M., A. Guarneros, and V. Nekrassov, *Effects of carbamazepine, phenytoin, valproic acid, oxcarbazepine, lamotrigine, topiramate and vinpocetine on the presynaptic Ca2+ channel-mediated release of [3H]glutamate: comparison with the Na+ channel-mediated release.* Neuropharmacology, 2007. 53(7): p. 854-62.

327. Sitges, M. and V. Nekrassov, *Vinpocetine selectively inhibits neurotransmitter release triggered by sodium channel activation.* Neurochem Res, 1999. 24(12): p. 1585-91.

328. Sitges, M. and V. Nekrassov, *Vinpocetine prevents 4-aminopyridine-induced changes in the EEG, the auditory brainstem responses and hearing.* Clin Neurophysiol, 2004. 115(12): p. 2711-7.

329.   Kaneko, S., et al., *Effects of several cerebroprotective drugs on NMDA channel function: evaluation using Xenopus oocytes and [3H]MK-801 binding.* Eur J Pharmacol, 1991. 207(2): p. 119-28.

330.   Krishtal, O., et al., *Modulation of ion channels in rat neurons by the constituents of Hypericum perforatum.* Pharmacopsychiatry, 2001. 34 Suppl 1: p. S74-82.

331.   Muller, W.E., A. Singer, and M. Wonnemann, *Hyperforin--antidepressant activity by a novel mechanism of action.* Pharmacopsychiatry, 2001. 34 Suppl 1: p. S98-102.

332.   Gleitz, J., A. Beile, and T. Peters, *(+/-)-Kavain inhibits veratridine-activated voltage-dependent Na(+)-channels in synaptosomes prepared from rat cerebral cortex.* Neuropharmacology, 1995. 34(9): p. 1133-8.

333.   Gleitz, J., et al., *Anticonvulsive action of (+/-)-kavain estimated from its properties on stimulated synaptosomes and Na+ channel receptor sites.* Eur J Pharmacol, 1996. 315(1): p. 89-97.

334.   Gleitz, J., et al., *Kavain inhibits non-stereospecifically veratridine-activated Na+ channels.* Planta Med, 1996. 62(6): p. 580-1.

335.   Gleitz, J., et al., *The protective action of tetrodotoxin and (+/-)-kavain on anaerobic glycolysis, ATP content and intracellular Na+ and Ca2+ of anoxic brain vesicles.* Neuropharmacology, 1996. 35(12): p. 1743-52.

336.   Magura, E.I., et al., *Kava extract ingredients, (+)-methysticin and (+/-)-kavain inhibit voltage-operated Na(+)-channels in rat CA1 hippocampal neurons.* Neuroscience, 1997. 81(2): p. 345-51.

337.   Schirrmacher, K., et al., *Effects of (+/-)-kavain on voltage-activated inward currents of dorsal root ganglion cells from neonatal rats.* Eur Neuropsychopharmacol, 1999. 9(1-2): p. 171-6.

338.   Omiya, Y., et al., *Antinociceptive effect of shakuyakukanzoto, a Kampo medicine, in diabetic mice.* J Pharmacol Sci, 2005. 99(4): p. 373-80.

339.   Suzuki, Y., et al., *Antinociceptive effect of Gosha-jinki-gan, a Kampo medicine, in streptozotocin-induced diabetic mice.* Jpn J Pharmacol, 1999. 79(2): p. 169-75.

www.ingramcontent.com/pod-product-compliance
Lightning Source LLC
Chambersburg PA
CBHW041709210326
41598CB00007B/588